RADA[illegible]
MATE

RADAR MATE

Lt. Cdr. G. A. G. Brooke
and
Capt. S. Dobell

ADLARD COLES LIMITED
8 Grafton Street, London W1

Adlard Coles Ltd
William Collins Sons & Co. Ltd
8 Grafton Street, London W1X 3LA

First published in Great Britain by
Adlard Coles Ltd 1986

Distributed in the United States of America
by Sheridan House, Inc.

British Library Cataloguing in Publication Data
Brooke, G.A.G.
Radar mate.
1. Radar in navigation 2. Boats and boating—
Radar equipment
I. Title II. Dobell, S.
623.89'33 VK560

ISBN 0-229-11789-9

Typeset by BookEns, Saffron Walden, Essex
Printed and bound in Great Britain by
R.J. Acford Ltd

Contents

Introduction

Radar Mate is primarily intended as a concise guide to radar which can be kept alongside the display for reference on practical points at any time. Some basic explanations are included to set the scene for the completely uninitiated, but radar for small craft is now so well established that this section is kept to the minimum. Those wishing to go into more actual detail should get one of the many books on the subject (such as the classic *The Use of Radar at Sea*, by Captain F. J. Wylie, published by Hollis & Carter).

Plate 1 Radar for small craft is now so well established – on the Hamble, 14 radars are visible

There is little doubt that many owners of radars do not get the most out of their considerable investment. Some may think they have not the time, and others, let's face it, simply do not have the inclination to explore its potential in any depth. So long as their Decca Yacht Navigator shows them where they are to within yards, and the radar indicates that something is there, or is not there, that's enough!

This approach is a pity. Not only is some study repaid by the provision of a greater margin of safety both for the user and for others, but it will also be found to open up a whole new avenue of interest. To promote this, a section on plotting (by which, for instance, the courses and speeds of radar echoes in bad visibility are determined for obvious benefit of the observer) is included.

It is assumed throughout that the radar used is a typical 'relative motion head-up' small craft set. Naturally these vary a little from manufacturer to manufacturer and where it is necessary to describe a particular function, the Racal-Decca 170, 270, 370 series of conventional radar has been taken.

Relative motion (RM) is the mode of presentation of radar data in which the position of own ship, and therefore the origin of the radar transmissions, is always at the centre of the display. Because all echoes are 'painted' by own ship's aerial it follows that they will be presented as relative to own ship (and their courses and speeds about the display will also be relative).

When echoes move with their actual courses and speeds, this is known as 'true motion' (TM), and is effected in big ship radars by moving the centre spot via inputs from the gyro compass and log. A few yachts world-wide may have TM radar, but this development has been virtually ignored here, as being likely to cause confusion. However, TM is touched on as a fact of maritime life, e.g. when the boat is stopped (and is heading True North), all echoes will be shown moving with their actual courses and speeds, that is, with true motion.

Where the word 'miles' appears in this book, it should be taken to mean 'nautical miles'. This colloquialism is often found in nautical books and is used by most seafarers.

Radar Mate is designed so that any page of the text can be laid open with its relevant diagram(s), where provided, alongside.

Basics

Some old hands may scoff at electronic navigational aids, as old fishermen did at the first fishing echo sounders, but radar is now so widespread that all those owners cannot be wrong. Just as sounders really do assist in catching fish, radar helps you to keep out of trouble, both from collision and grounding – its *raison d'être* being translatable as 'peace of mind'. Yachting is, after all, supposed to be a leisure activity and the fun can go out of it after a few hours of straining through a curtain of darkness or bad weather.

Though it helps, you do *not* have to know what goes on behind the picture or under the revolving aerial (scanner), provided you have taken in the salient points on operating it from the sales engineer concerned, reinforced by the instructional manual (and reminded by *Radar Mate*!). But it is important to appreciate that a radar is not just a second cousin to a TV; it *does* need trained servicing (from the manufacturer's authorised dealers, wherever possible).

Plate 2 Bracket mounting of an RD 370 scanner

How does the thing work? It is easy to understand if you take the analogy of a man shouting towards a cliff face or other good reflector of sound. The speed of sound is constant, and so if you start a stopwatch at the moment of 'transmission' and stop it on receipt of the echo, it is easy to work out how far you are from the cliff. Radio waves transmitted with enough power and at very high frequency bounce back from hard objects in exactly the same way. Their speed is also constant, so the next and logical step is to provide mechanical display of range and bearing.

The technical term for a radar picture is PPI (Plan Position Indicator). The observer is looking down on the scene as if from a helicopter flying so as to remain *always immediately above* your own boat, shown as a point of light in the middle of the display. The ship's head is indicated by the heading marker, an electronic line that always extends from the centre spot to the top of the display. Fitted over the PPI is a revolving Perspex cursor with a bearing line engraved on it and the 360° of the compass circle (see Fig. 1). To read off a bearing, you move the cursor until the bearing line cuts the echo and note where the heading marker intersects the bearing scale. Alternatively, there may be an electronic bearing line (EBL) which is rotated, the bearing being indicated by digits. The result is a *relative* bearing, i.e. relative to your ship's head, and the whole business is known as 'relative motion head-up'.

Range is determined by concentric rings at intervals. At the touch of a control, the radius of the display can be made to represent the radar's full range (usually 16, 24 or 48 miles, according to the size of the set), or a number of shorter distances down to ¼ mile or so. Unless the echo happens to be exactly on a ring you interpolate by eye, but for greater accuracy use the variable range marker (VRM), if fitted. With a VRM there is a variable range ring which is expanded or contracted to cut the echo, the range being shown in an adjacent window.

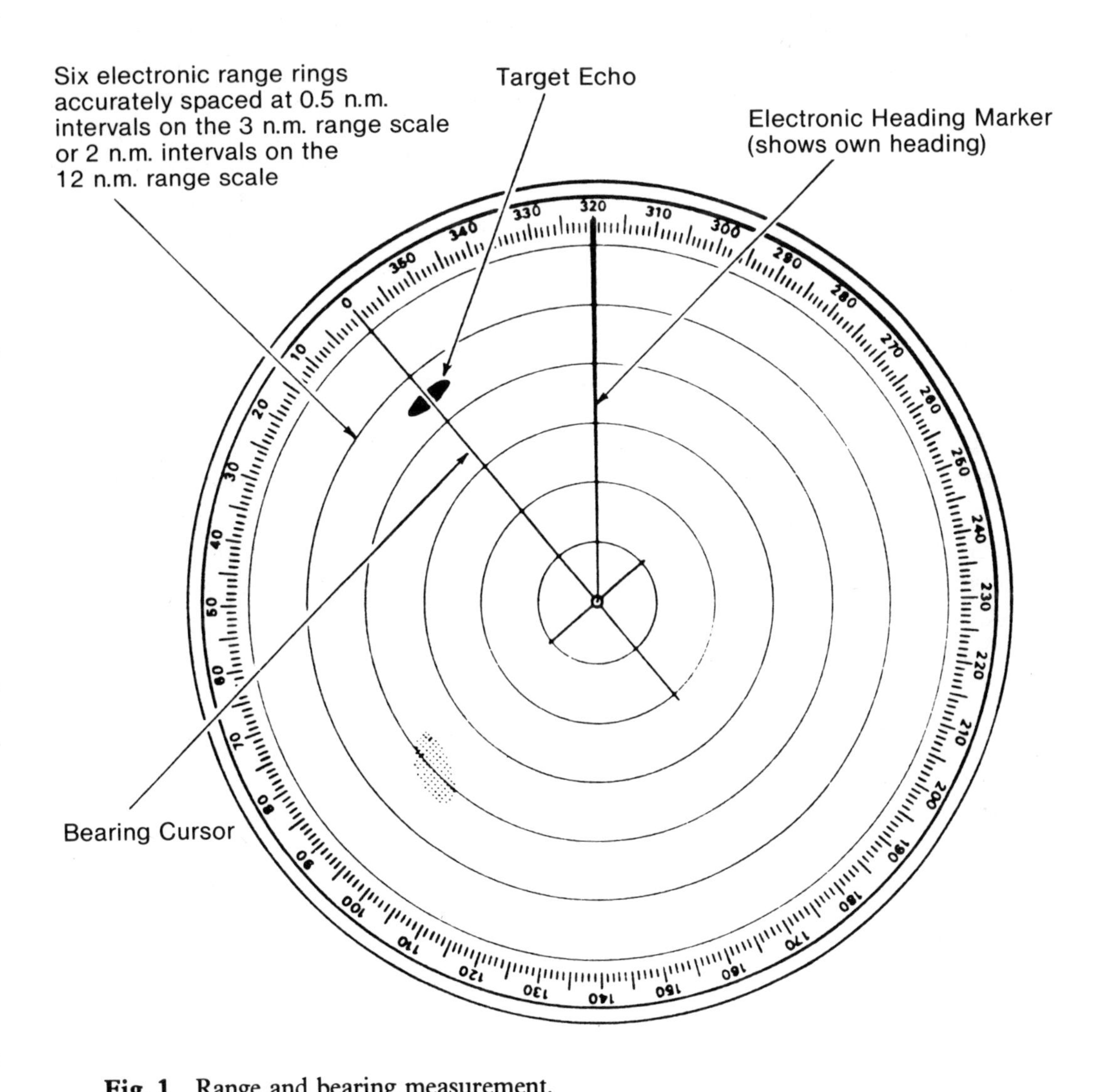

Fig. 1 Range and bearing measurement.

This measurement of range and bearing is easy to describe but somewhat laborious to bring about, which accounts for most of the expense if it is to be done accurately and reliably. Briefly, the original transmission is punched out by a valve called the magnetron and radiated horizontally by a rotating scanner. This also picks up the returning echo and passes it down to a receiver where the echo is processed and passed to the display, the end of a cathode ray tube (somewhat similar to that of a TV). The position of the echo on the screen in azimuth (bearing) is achieved by synchronising the painting of the echo with the rotation of the aerial, and the range by measuring the time intervals (based on the speed of radio wave travel) to determine distance from the centre of the screen. The transmitter and the receiver are combined into one unit known as the transceiver and there has to be a power unit so the radar can operate on the boat's electricity supply. Including the scanner, there are thus five components in a radar system. For convenience they are, in yacht radars, grouped into only two units, the scanner assembly (or top unit) and the display, which contains the rest.

For a more detailed description of the working of a radar turn to Appendix I, and for a glossary of terms, Appendix II. A study of these is recommended if you are to get full value from what follows.

Operational Considerations

The siting of your radar – both scanner and display – is very much a matter of taking the dealer's advice, and normally, of letting him do the installation. Though at least one radar is advertised as DIY, this is not recommended as few circumstances are identical. The higher the scanner the better for range purposes, but this conflicts with both the topweight consideration and the fact that sea clutter increases with scanner height; also, it is a mistake to put the scanner out of reach of a bos'n's chair as this may necessitate striking the mast for servicing. Locating the display should be done with due heed to mutual electronic interference and compass safe distance. Scanners can be mizzen-, canopy-, bracket- or pedestal-mounted; displays can be table-top, bulkhead, deckhead or flush in a console.

No reputable sales engineer will sell a radar that is too big for the boat. But this said, it is a fact that a prospective buyer is well advised to go for the largest radar that his pocket, boat size and available power will stand. In this connection there are four basic considerations: maximum range, scanner length, display size and display type.

Plate 3 Powerful dual radar installation on this luxury motor-cruiser consists of RM 914 (upper) and RM 916. Big globe is a SATCOM aerial

Range

Maximum range is basically a factor of transmitted power, receiver efficiency and scanner size, so obviously the larger the radar the longer its potential range. 'Potential' because radar sees almost directly, like the human eye; its transmissions cannot travel through solids or far over the horizon like radio. The horizon of a radar (the range at which it will pick up a given target) is the radar's own horizon, which varies with its height, plus the similar horizon of the target.

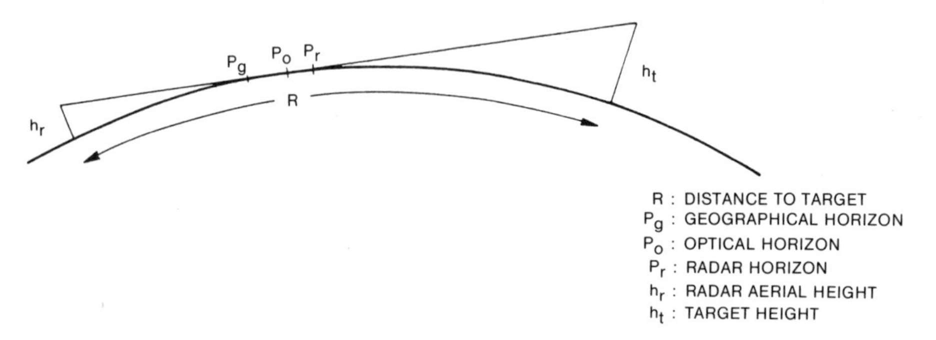

Fig. 2 True target distance using radar horizon.

To take an example, a radar at 20 ft will see a target of 35 ft at 5½ + 7 = 12½ miles (see Appendix III for more detail and examples). Therefore, if maximum range is the only parameter of interest, there is no point in fitting a 48-mile radar that can only see a small vessel at 12½ miles. However, this is not the whole story; a 48-mile radar will have other qualities or facilities not available with its little brother, which may be just what you want – often the most obvious is particularly good short range performance, ideal for, say, picking up your buoy in fog. Many work-boats, like pilot launches, have large radars for this sort of reason. (Thames fire tenders have 48-mile radars, but it is doubtful whether they ever use a range scale above 3 miles.) And it stands to reason that a larger, and therefore longer range radar will have more comprehensive facilities for coping with the problems mentioned later.

A point to remember is that you should not be fooled by the maximum range scale on some radars, which, like car speedometers, suggest more than they can deliver.

Scanner length

This is a basic parameter in radar design that, unlike some others, cannot be side-stepped in any way. Greater scanner length allows the outgoing pulses to be more concentrated, which has the effect of greatly increasing radiated power (a 4 ft scanner has nearly twice the effect of a 3 ft one). A good analogy is a garden hose: with a spray nozzle you cover a wide area, but with a jet nozzle the water goes much further and can hit individual flowers. A small aerial is equivalent to a spray, but a large, jet-like aerial not only gives greater range but also narrower beamwidth, which leads to better bearing accuracy and bearing discrimination (the ability to separate adjacent echoes). Naturally, a small radar cannot so easily support a big scanner.

The smallest radars, of which there are a considerable number in competition, have their aerials inside fibreglass 'radomes'. Radomes reduce cost and power needed by allowing lighter aerials and smaller turning motors (wind resistance being virtually obviated). There is also the slight advantage for sailing yachts of not fouling running rigging.

Plate 4 Typical motor-sailer with mizzen installation. Note radar reflector

Display size

Obviously the bigger your radar picture the better (though 16 in. is considered the maximum that can be readily assimilated) in order to facilitate interpretation of what is going on around you.

Display type

There would be a serious omission in *Radar Mate* – becoming increasingly important as time goes on – if it was not pointed out that a new type of display has appeared that is superior to the conventional one in every respect, except price. This is the digital scan conversion (DSC) daylight viewing display, which is usually in colour. Contrary to traditional practice, it appeared first in small size (for technical reasons) but the first large colour display (the Racal-Decca 2090 BT) is now on the market, and the type will eventually be in general use in all large ships as well as yachts and work-boats.

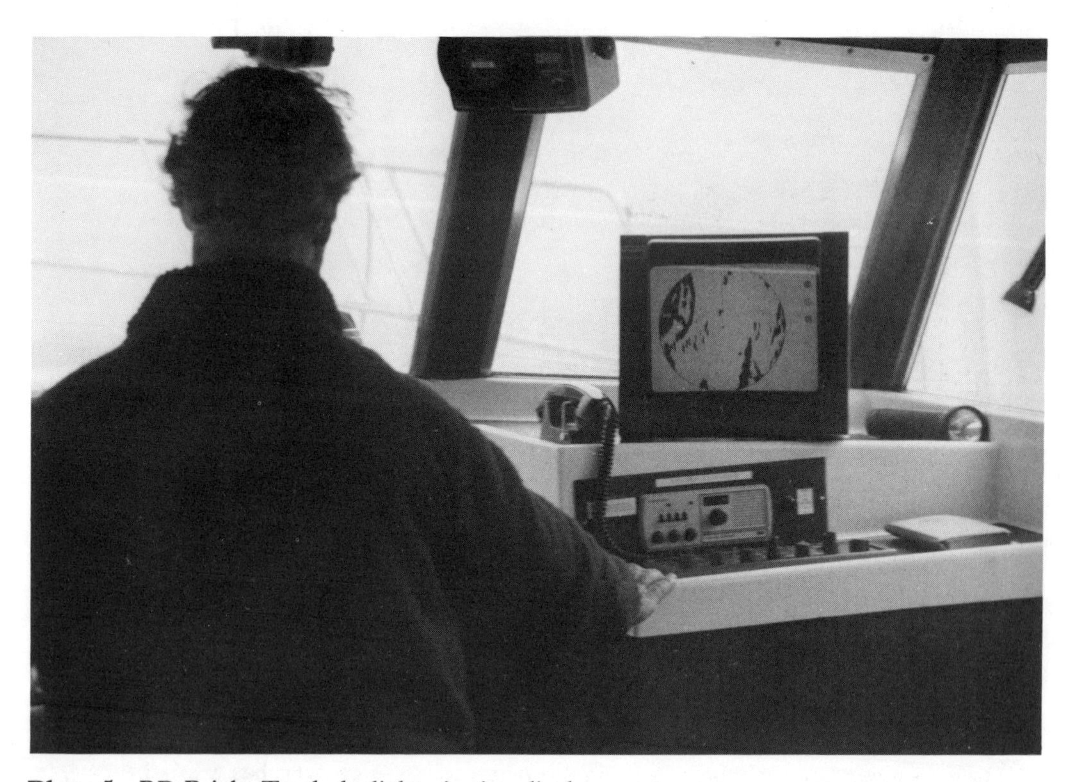

Plate 5 RD Bright Track daylight viewing display

The advantages of DSC are difficult to put in any order but the four main ones are:

(a) Tremendous brightness. Gone are the days of going right up to a radar and peering into its visor; it is best seen from, say, five ft, and can be viewed by several people at once from ten.
(b) Steady picture. There is no revolving segment of renewed echoes. Not only does this facilitate study of a feature, but it is also easier on the eyes.
(c) In one example, the five colours greatly help to distinguish between echoes, tracks, range rings, etc. (Although there is a choice of blue or black background colour for day or night, echoes in yellow are common to both – maintaining continuity and a link with conventional displays – ancillary information appearing in red, white, green or black.)
(d) Relative tracks of echoes are shown. Not only do these indicate collision risk, but being electronically generated, they are proportional to speed, a big advantage; with practice, true speed is not difficult to estimate.

For other advantages and a longer description of how DSC works see Appendix IV.

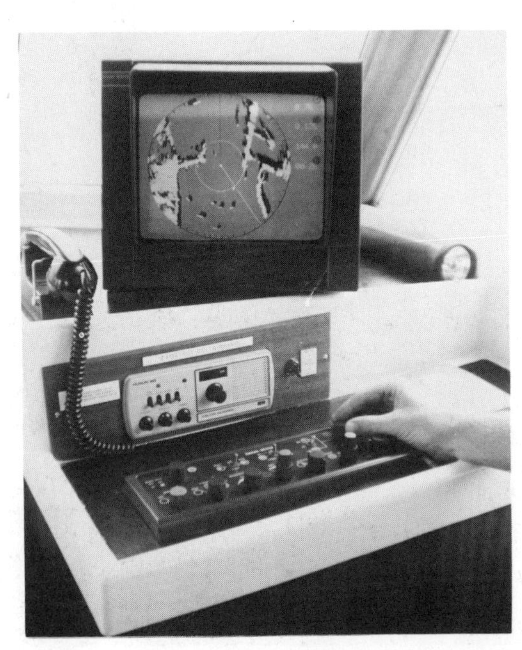

Plate 6 Colour display. Black shadows are target tracks. Land also has tracks because, to the radar, it has relative movement

We now come to uses of radar, interpretation of the picture, and several common problems.

Collision avoidance

This was the first object of commercial radar and is still its primary role in the minds of many. The time-honoured dictum that if the compass bearing of a closing vessel remains substantially unchanged she is on a collision course, can be put into practice at least as easily with a radar as with a bearing compass. Place the bearing cursor on the echo in question and note the bearing (and the ship's head, see below). Repeat a few minutes later and determine whether the bearing has 'grown' forward, aft, or remained steady. (See Fig. 3.)

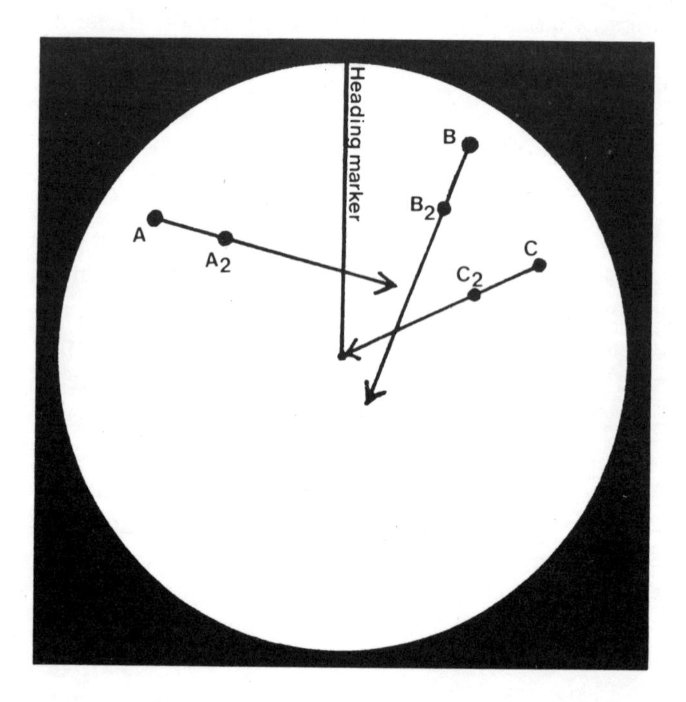

Fig. 3 Collision threats – A will pass ahead, B will pass astern, C is on a collision course.

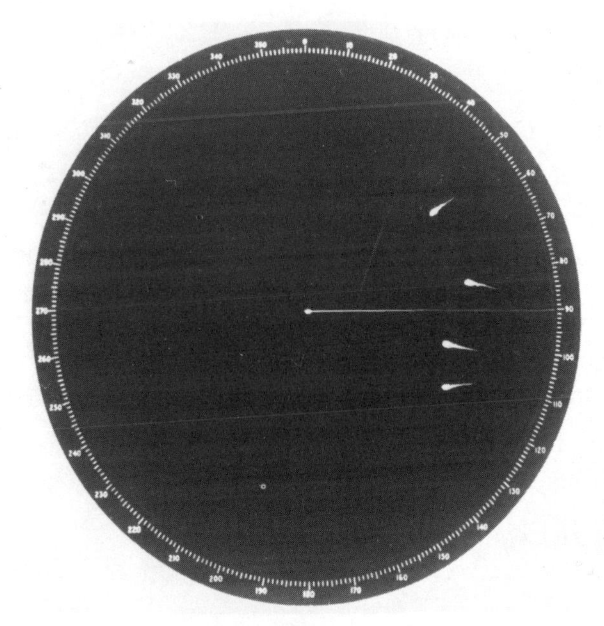

Plate 7 Echo trails, known as 'tails'

The tails that stretch behind echoes are useful as a guide to collision threat. (With conventional radars, now being considered, they are caused by the afterglow of previous 'paints'.) They do not, as one might expect, represent the course of the echo but its course relative to the centre spot (since all movement on our type of radar is relative). This means that if the tail projected forward cuts or passes near the centre spot (own ship), the echo is on a collision course. If this line misses to one side, the echo will pass ahead or astern as indicated. Tails painted in this way are not long and so are plainly visible on short ranges only. A rough idea of relative speed may also be obtained as the higher the speed the longer the tail, but this may only be feasible with something like a hovercraft. It must be remembered with relative motion and your boat under way, that stationary objects such as ships stopped, buoys, and land will have tails because to your radar they are closing or opening and so appear to be on the move. A quick survey of all the tails on a radar picture will tell at once which need special attention as collision threats.

Another collision avoidance dictum goes that it is irresponsible in conditions of nil or bad visibility to make an alteration of course without first ascertaining the other ship's course and speed. This presupposes an inclination and ability to plot, i.e. mainly using relative radar data to ascertain the other ship's true movements. Plotting is discussed in detail later.

Rule of the road

The international *Rules of the Road at Sea* (obtainable as a booklet from HMSO) should be understood by all seagoers, down to the smallest boat. Apart from the rule requiring the correct navigation lights to be shown in reduced visiblity, the following rules are particularly relevant:

(5) Look-out by sight, hearing, radar, VHF. Though radar is probably the principal navaid concerned, full use should be made of whatever else is fitted, such as Decca Yacht Navigator, echo sounder, DF or Satnav.
    Remember that these are only *aids*; none of them precludes the taking of normal seamanlike precautions, as laid down in sections, 6, 19, 34 and 35 which follow:

(6) Safe speed. This refers to slowing down (and if necessary stopping) in bad visibility.

(19) Conduct of vessels in restricted visibility.

(34) Manoeuvring and warning signals by vessels in sight of one another.

(35) Sound signals in restricted visibility.

The last two may not apply to *you*, but it is important to know what the other ship means by her signals.

Three other points are worth mentioning before leaving collision avoidance:

(a) In the interests of a clear, unambiguous display with 'punchy' echoes, a useful option (for the radars mentioned in the Introduction it is called VP (Video Processing) 3) provides for automatic suppression of interference (see page 26), for the brightening of weak echoes, and the enlarging of distant ones. VP3 can be switched in or out.

(b) Transmitted pulses are lengthened when range scale is increased. This can cause near but poorly reflecting objects to be missed, so you should switch over to a short range from time to time in order to 'have a good look round'.

(c) Always wear a good radar reflector so that your boat shows up on other vessels' displays.

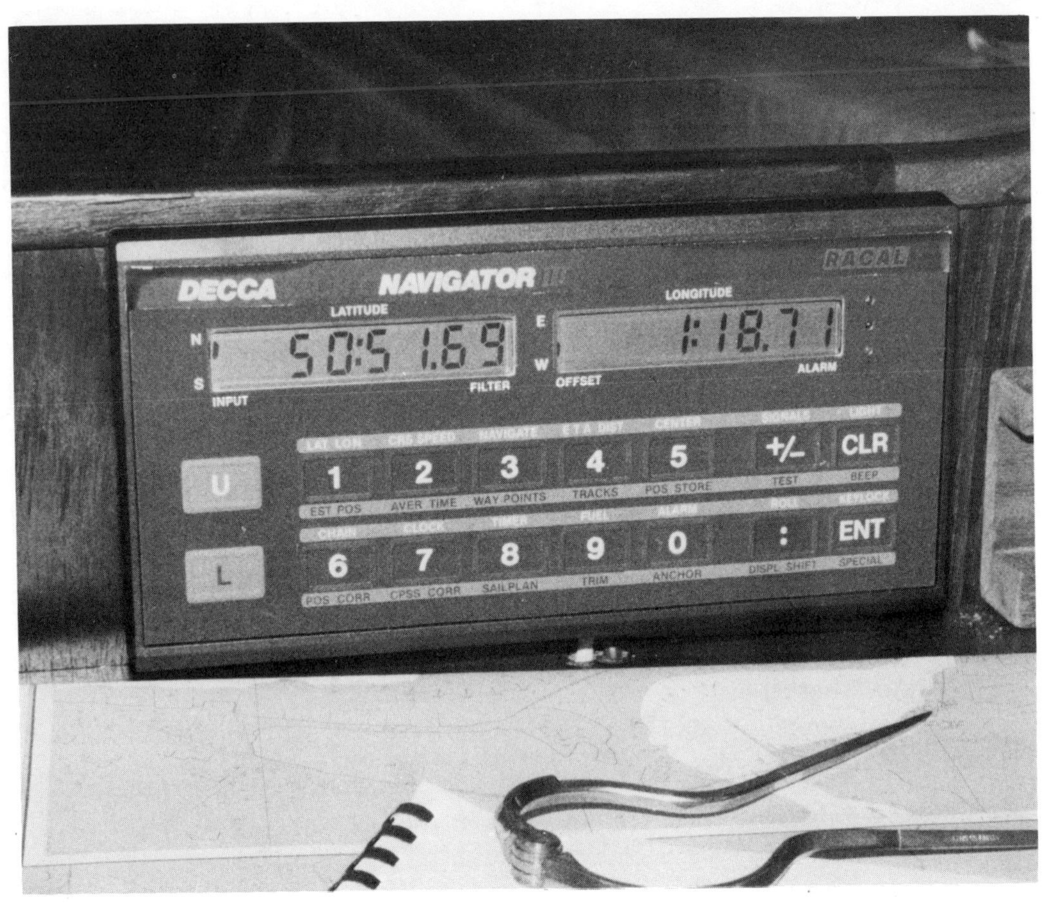

Plate 8 Decca Yacht Navigator III

Points on measuring range and bearing

Range: The leading edge of the echo (i.e. the part nearest the centre of the screen) should be taken.
Bearing: The bearing cursor or EBL should be made to cut the centre of the echo at the moment when the rotating trace is crossing the echo (so ensuring there is no parallax involved). Radar's tendency to stretch out an echo has the result of elongating headlands. To allow for this the bearing should be corrected towards the land by half a beamwidth of the radar, say 1°, but see specification of your set.

Remember that whenever taking bearings in a normal yacht radar, they are relative to the ship's head. If she happens to be dead on course when the bearing is taken, all well and good, but if she is yawing about, the bearing will have to be corrected for the amount off course at the time. When in suitable circumstances in calm weather, it is possible to obviate this calculation by waiting until the boat is dead on. However, this is bad practice. There will be occasions when, for instance, you will be taking a three-point fix and delay is unacceptable, or you will want to use 3- or 6-minute intervals to facilitate speed calculations (6 being 1/10 hour). It is therefore better always to do the calculations, so that they become second nature.

A rule-of-thumb method of changing radar bearings to relative bearings is:

(a) Ship's head is noted – if possible by an assistant – each time bearing is taken.
(b) Course error is then written down. This is the number of degrees the boat is off course, with + for an above-course figure and – for below.
(c) Apply course error to radar bearing.

Examples for course of 096°:

Radar bearing	Ship's head	Course error	Relative bearing
013	096	0	013
356	097	+1	357
057	099	+3	060
271	093	–3	268

Plate 9 Typical mizzen and canopy installations

Position fixing

Used intelligently (and this is where 'getting the most out of your radar' comes in), radar is of equal if not greater use for position fixing (and plotting) than for collision avoidance. In fact, during one cruise when a careful record was kept by an expert, radar was employed for collision avoidance only one to every three times for the other reasons. In coastal navigation the usual position fixing ploys can be carried out provided two points are borne in mind.

First, though the range facility is very accurate, bearing is not so accurate (error being from $\pm 1^\circ$ to $\pm 3^\circ$, depending on the quality of the radar – another good reason why a big 'un is better than a little 'un). The choice of methods in order of preference is therefore: radar ranges and compass bearings; two, or preferably three, radar ranges; radar ranges and radar bearings a poor third; and radar bearings alone as a last resort. 'Distance off' can often be employed.

To fix by ranges: measure the radar ranges of the points chosen and draw to scale on the chart corresponding arcs from their positions. The three arcs should coincide at your position. (See Fig. 4.)

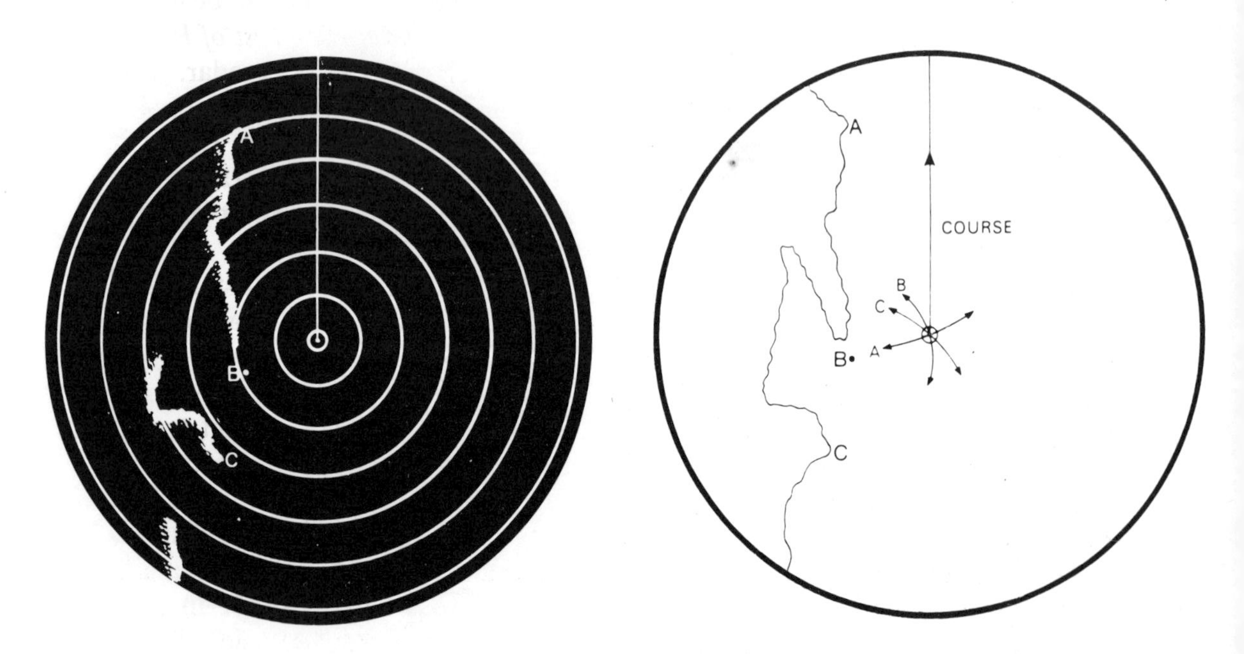

Fig. 4 Position fix by radar on three marks.

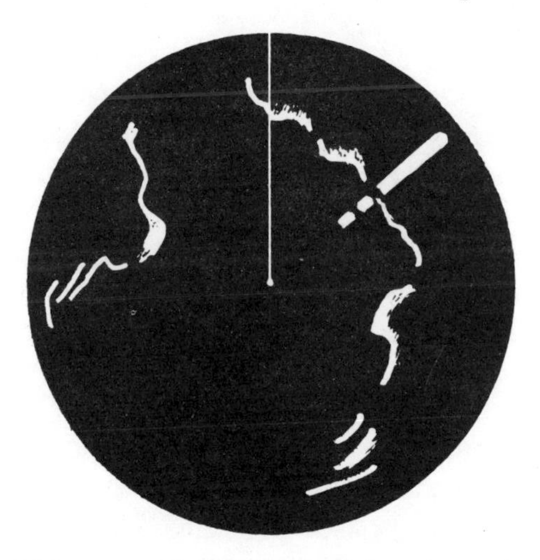

Fig. 5 Signal on ship's PPI as a result of its radar triggering a Racon.

Racons provide a navigational facility which is both unique to radar and becoming more important as stations proliferate. Racons are Radar beacons, situated along frequented coasts or on lightvessels (see *Admiralty List of Radar Signals* Vol. 2), which emit an identifiable signal when triggered by a radar. This response on your display appears as a radial line or broken line, its nearest point being the location of the station (see Fig. 5). (Some, called Ramarks, transmit independently and provide bearing only.) Transfer of the Racon's range and bearing to the chart will of course give position. Some radars cannot pick up Racons and these should normally be avoided.

When relative bearings (as above) are taken of points ashore for transfer to the chart, they must of course be corrected for (a) ship's head, (b) deviation, and (c) variation.

Radar is not impeded by nightfall or bad visibility, but is little use for position fixing if – and this is the second point – you cannot identify your target on the chart. It is always necessary to underline this – a matter which does need practice. It can happen to the best of us that, when transferring a range or bearing to the chart, the wrong point is taken. No difficulty should arise in the case of a very 'conspic' object such as the end of a breakwater, or lone building on top of a hill, but the poor reflecting qualities of some features may cause trouble in that, though you can see the land quite clearly with the eye, it may, occasionally, be returning no echo. In which connection you are strongly advised to get to know really well all the useful radar marks (by day, and any lights at night) around your home and other frequented ports. Which leads to . . .

Interpretation of the display

This is a subject in itself, but comes easily enough with practice. In fact, it cannot be too strongly stressed that you should practise this (and in fact all radar usages) in good visibility when mistakes may not matter, so that there is complete confidence when conditions are bad.

It helps to know what produces good, and what poor echoes. First, the reflecting properties of a target vary with size, shape, material, texture, and aspect. Obviously, a supertanker shows up better than a tug; a flat surface is better than a sphere or cylinder; steel is better than wood or grp and soft materials are bad; a rough surface is better than a smooth; and lastly, a surface at right angles to the transmission is better than one at an oblique angle. So, on occasions a clear enough object may be providing little return. Usually this applies to something like mudflats (soft, smooth, bad aspect) and is the reason why that nice, clean echo line is really coming from a sea wall beyond and if you don't watch out you will soon be on the putty!

When looking at land on a display it is necessary to remember that radar has no X-ray facility to go *through* a feature as radio does, so that certain blank spaces do not necessarily indicate that they contain nothing that reflects; the area is simply blanketed by higher ground in front. Viewed from further along the coast the same terrain may give somewhat different returns.

It should be remembered that a small wooden or grp boat is usually detected by a large ship's radar at between ½ and 5 miles, whereas the same radar will pick up a big ship target up to 20 miles off. Again, always wear a radar reflector.

Radar log

It is worth giving serious consideration to keeping a radar log. Apart from the record of hours run, faults, control settings found to give best results under various conditions, etc., an 'Observations Made' column can be of immense help if kept up, e.g. 'Cap Prince cliffs seen at 12 miles, but lighthouse at only 2 miles' ... 'very small wooden fishing boats used in area, which do not show up on screen over 1 mile. Keep switching to 3-mile range to change pulse length'. Also, local dealers' details, useful telephone numbers, etc., can be included.

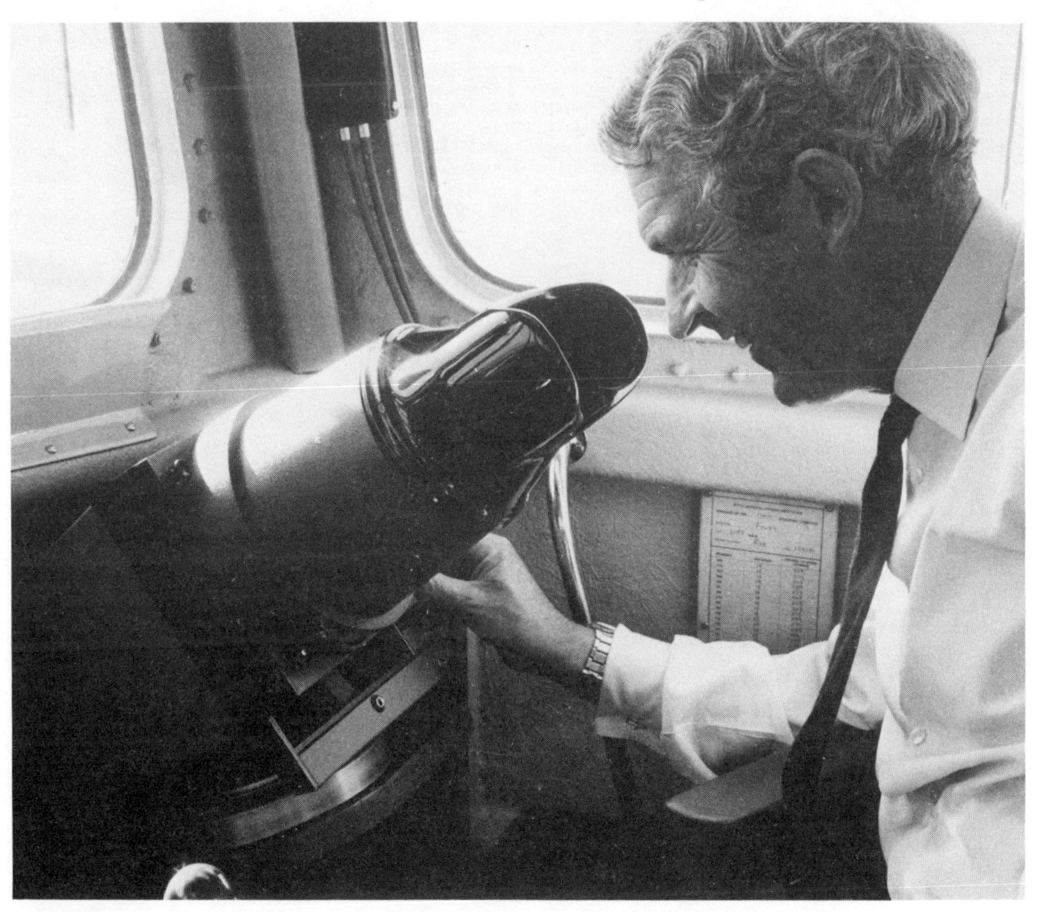

Plate 10 Admiral Graham, Director, RNLI, with a table-top RD 370 display in a lifeboat

The be-all and end-all of a yacht radar is to provide a clear picture, which is easily understood without effort, with simple measurement of range and bearing. There are some natural and some man-made enemies of this ideal, the worst of which are sea clutter, rain clutter, and interference from other radars.

Sea clutter

The waves of a choppy sea return echoes, producing a scintillating effect (since the waves are never in exactly the same place twice) out to about two miles from the boat (Fig. 6). The effect will be worse, and extend further, to windward because of the better reflecting qualities (from steeper surfaces) in that direction.

Sea clutter is dangerous in that it tends to obscure 'legitimate' echoes from ships, etc. There is an Anti-Clutter (sea) control and this and the gain control should be adjusted in concert. The gain control should normally be set so that the speckled background is just visible at longer ranges. A common fault is to apply too much suppression with the Anti-Clutter (sea) control – a blank area may mean that legitimate echoes have been suppressed too. It is preferable for the adjustment to be correct at the outer extremity of the clutter zone, some saturation of the close ranges being accepted. If after adjustment of the A/C control there are dark patches among the sea clutter echoes, it means that in these places the gain is too low. The observer should return to the display and 'play' with both controls from time to time.

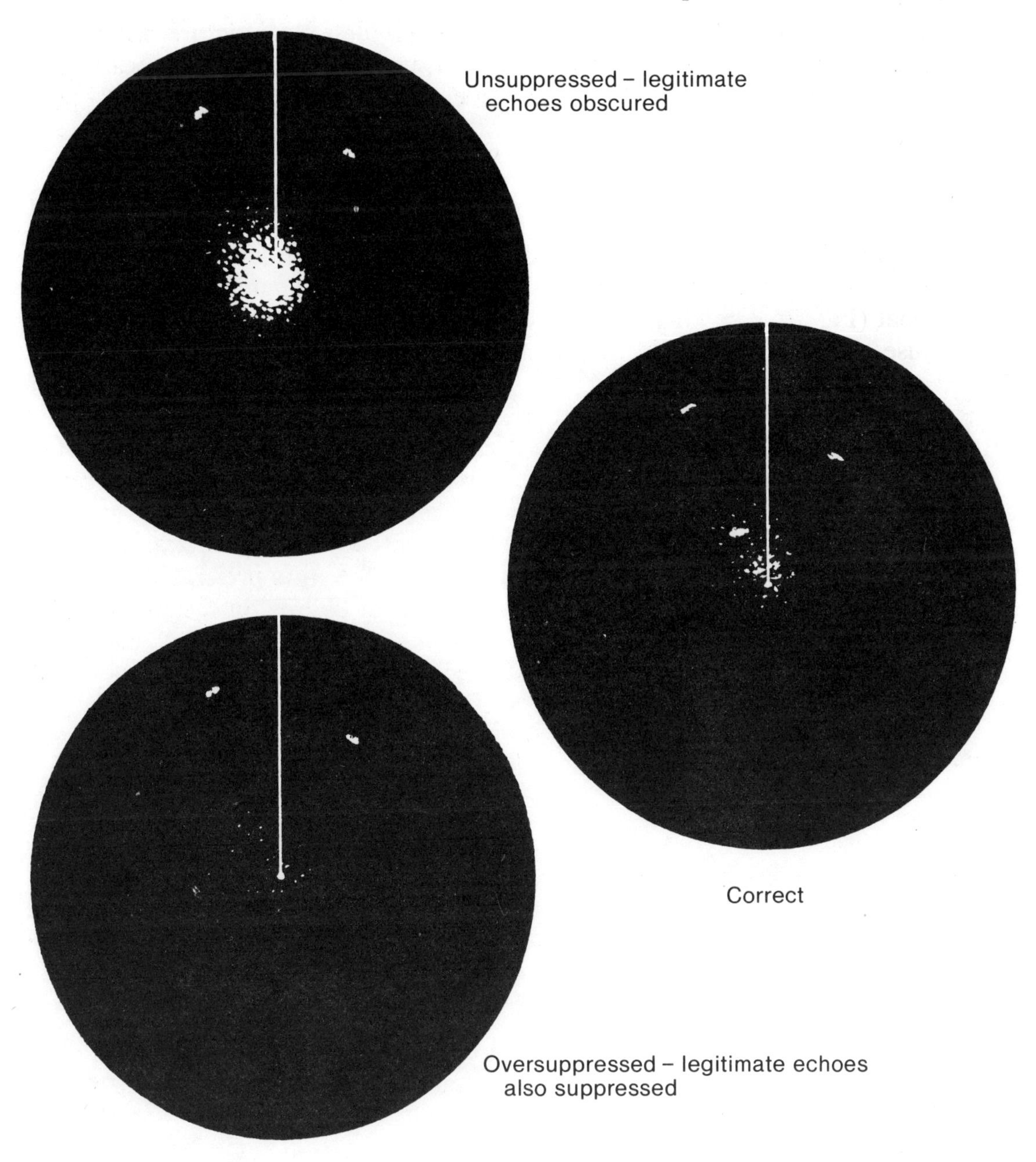

Fig. 6 Use of Anti-Clutter (sea) control to suppress sea clutter.

Rain clutter

A very new owner tells the story against himself that he made a wide detour at night round an island that was providing a huge echo. There was no sign of it on the chart and when it suddenly faded away he realised he had been avoiding a rain-cloud!

Raindrops and some other forms of precipitation such as snow (but generally not fog) return echoes that usually form a cotton wool effect (Fig. 7) on the display (though sometimes more solid) at, of course, whatever position the rain happens to be. Like sea clutter, 'rain clutter' is dangerous in that it tends to obscure legitimate echoes from ships, etc.

To overcome this as far as possible, gain should be reduced until the rain clutter is just visible, where legitimate echoes should show. In addition, or rather in co-operation with the gain control, the A/C (rain) control (also called differentiator – 'diff' – the name of the circuit concerned) should be adjusted so that the optimum level of both is arrived at. Remember that too much suppression of rain echoes may obscure the legitimate echoes you need to see, and it is wise to adjust the controls from time to time while studying the area concerned.

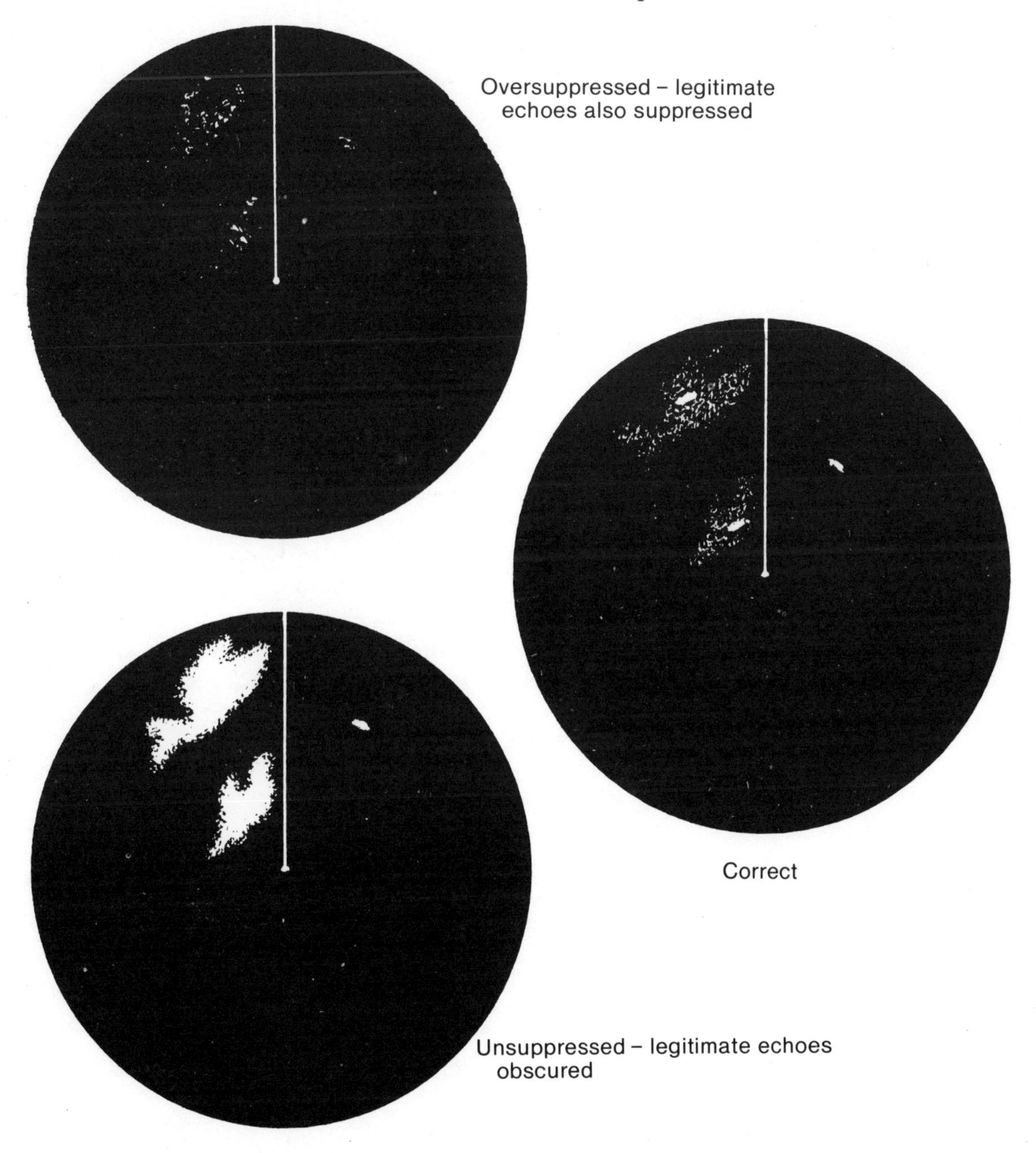

Fig. 7 Use of Anti-Clutter (rain) control to suppress rain clutter.

Interference from other radars

It happens that if another radar is operating close to you, on the same, or more likely a slightly different frequency, your display will show transient echoes in the form of dots in radiating curves or spirals. This can be a considerable nuisance and most modern yacht radars can be equipped with an interference suppression circuit to combat the phenomenon.

Other unwanted returns

There are quite a number of minor adverse effects that may be encountered from time to time, including:

(a) *Sidelobes.* As with a searchlight, it is not possible to concentrate all the radiated energy into a single narrow beam and some of it escapes to the sides. These 'sidelobes' are only strong enough to register as echoes at close range. Appearing at the same range as, but either side of the main echo, they can usually be eliminated by adjustment of the gain and anti-clutter controls.

(b) *Shadow sector(s).* As we know, radar cannot see through solid objects, so, for instance, the foremast of a boat – particularly if the radar is on the canopy – may produce a 'shadow' sector on the PPI. When this is the case the helm should be put over occasionally to uncover the blanketed area.

(c) *Multiple echoes.* If another vessel is passing at very close range, a second echo may be produced (beyond the normal one and at twice the range) as the original echo is reflected back and forth between the two vessels. There may even be a series of such echoes.

(d) *False or indirect echoes.* Due mainly to the wide vertical width of the radar beam, it is possible for returning echoes to be reflected off parts of own ship's structure, adjacent buildings, dock walls, etc. (in the case of small craft, usually only the last two). False echoes appear on the bearing of the obstruction but at the range of the legitimate echo, and can cause confusion unless very distorted in shape.

(e) *Spoking.* As implied, spoking appears in the form of lines radiating from the centre of the display. It is different from interference in that the lines are straight and not formed by a succession of pips. Spoking indicates a fault in the display such as dirty contacts on the bearing transmission.

(f) *Distortion.* On switching on, the picture may not fill the whole of the PPI. This is due to a build-up of static electricity inside the CRT, and should right itself quickly.

Other effects are mentioned briefly in Appendix II and can be read up in *The Use of Radar at Sea.*

Abnormal atmospheric conditions

Abnormal atmospheric conditions can produce the effects of super-refraction, sub-refraction, or ducting. The first of these occurs in calm conditions when there is a layer of warm air over a layer of cold, moist air; the effect is to increase the downward bending of the radar beam and *increase* range.

Sub-refraction occurs when the two layers are reversed, causing a *decrease* in range. It can also affect performance at short ranges, in that low lying targets may be missed.

Ducting is the name given to extreme cases of super-refraction, when the radar beam gets reflected upwards and downwards as if within a 'duct', resulting in exceptionally *long range.* Ducting may be encountered in the summer around the British Isles, often in the Mediterranean and elsewhere.

Plate 11 Pedestal mountings of two RD 370 BT colour radars on a motor fishing vessel

Switching on

For a description of the controls, see your radar handbook. Some radars will not have a 'Stand By' position.

(a) Ensure that correct power is connected to the radar (especially in the case of shore supply). If you have an open, rotating scanner, check that it is unobstructed. See that the mains switch of the boat is ON. Set the radar power function switch to the 'Stand By' position.

(b) Turn all relevant controls fully anti-clockwise (e.g. tube brilliance, gain, anti-sea and anti-rain clutter). Set radar range between 1 and 3 miles.

(c) Ensure that the ready lamp lights before proceeding further. (Approximately 3 minutes should elapse between switching to Stand By and the ready lamp lighting.)

(d) *Power/function switch* Set the switch to the on position. This starts radar transmission. The radar automatically changes to long pulse transmission from short pulse transmission at the 1.5 miles or the 3 miles range scale (depending on preselection at installation).

(e) *Panel lights control* Adjust the control to obtain optimum illumination of the front panel and bearing cursor plate.

(f) *Tube brilliance control* Rotate the knob gradually clockwise until the sweep trace becomes plainly visible. Rotate the knob anti-clockwise until it is only just visible (see Fig. 8).

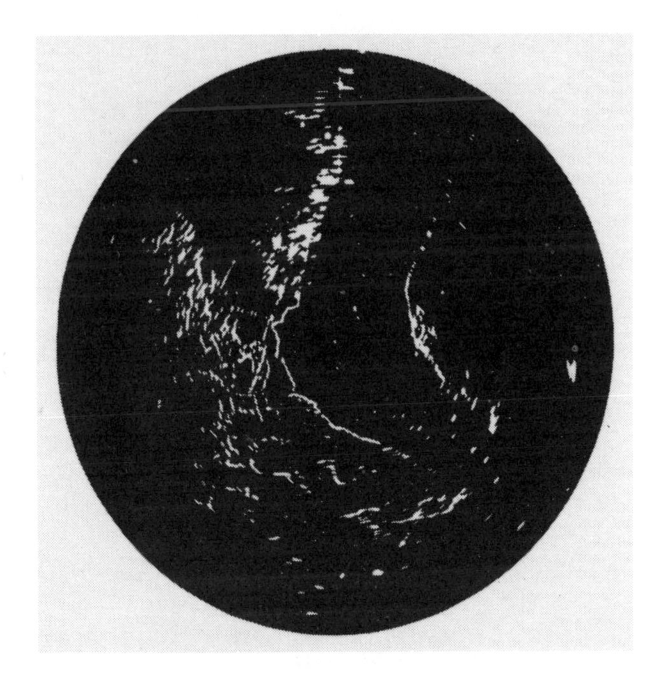

Normal Brilliance

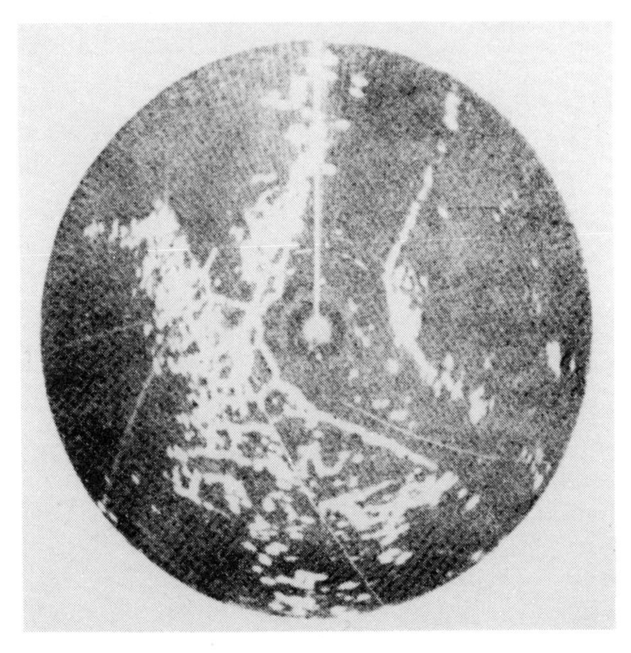

Excessive Brilliance

Fig. 8 Brilliance settings.

(g) *Gain control* Rotate the knob clockwise until receiver noise speckle appears faintly on the screen (see Fig. 9).

(h) *Range selector switch* Set the switch to the desired range, with targets clearly visible.

(i) *Tune control* For optimum tuning, adjust the knob for maximum possible brightness of the tuning indicator. It is advisable to make final adjustment visually, using the quality of picture on the screen.
NB Still further adjustment of gain and brilliance controls may be necessary to obtain the best picture.

(j) *Range rings On/Off switch* Switch on range rings.

(k) *Anti-Clutter (sea) control* Rotate the knob gradually clockwise until the effect of sea clutter, ground return or sidelobes has been reduced to the point where only minimal sea clutter remains visible and produces considerable reduction in gain when fully clockwise. It can also be used for short range gain control. See also page 22.

(l) *Anti-Clutter (rain)* Set the power/function switch to the 'A/C rain' position to break up (differentiate) into well-defined particle-like patterns the rain clutter caused by rain, snow or heavy ground returns. See also page 24.

(m) *Line up heading-marker 'head-up'* Consult *Owner's Manual* as to method. The HM is activated by a photo sensor each time the aerial crosses the dead-ahead position. It can cover a small echo, for which contingency a temporary OFF switch is provided.

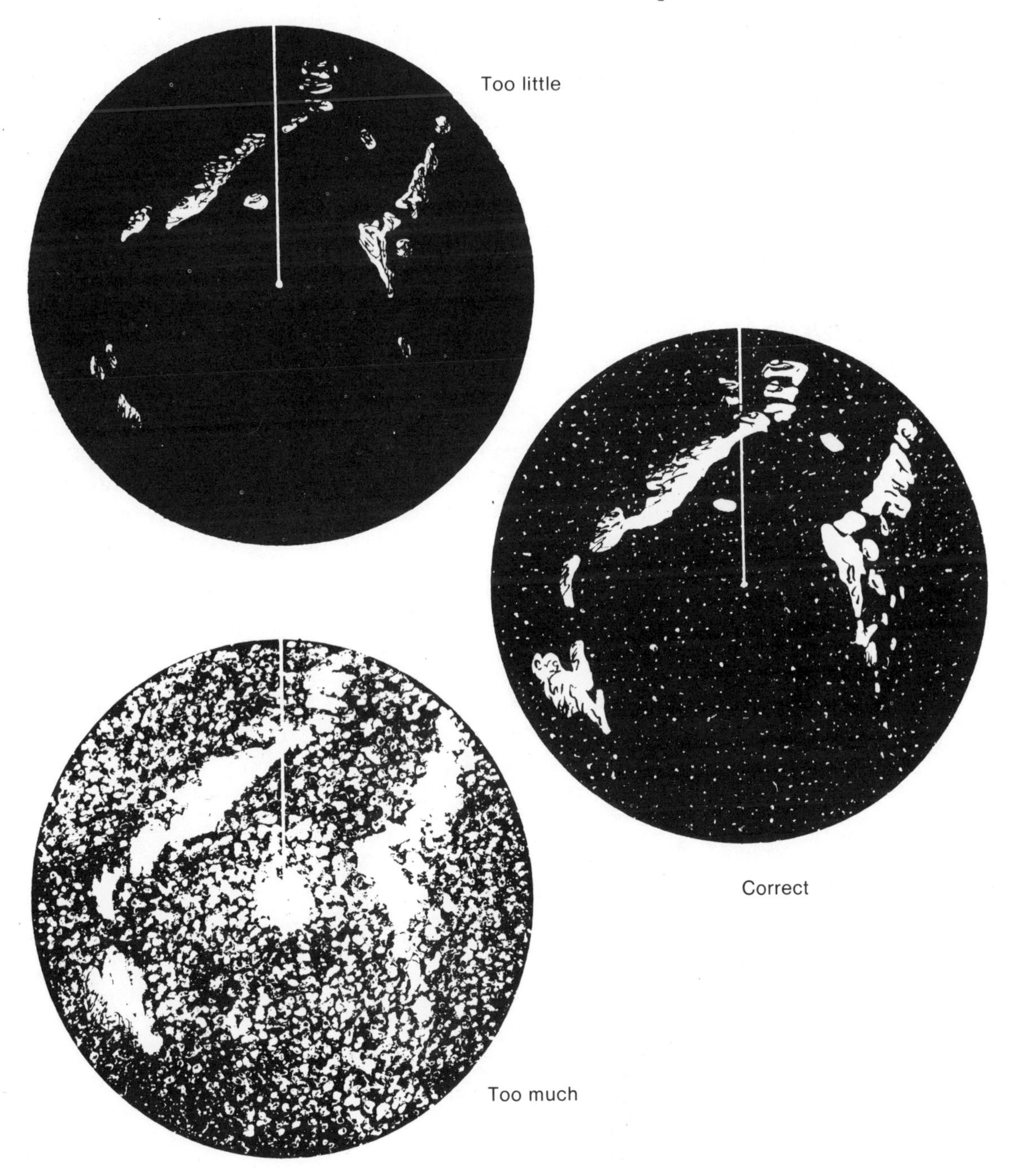

Fig. 9 Gain settings.

Normal range scale

Once tuned, if the radar is being used by more than one person, agree on the range to be normally used. This will avoid an error due to misunderstanding of the range in use. (Take an echo at 2 miles on the 3-mile range scale. To an operator who had previously left the radar on the 6-mile scale, this would appear to be at about 4 miles – a thoroughly safe distance!)

Switching off

(a) Rotate the tube brilliance and gain control fully anti-clockwise.
(b) Set the power/function switch to the 'Stand By' position to temporarily switch off; or to the OFF position to completely switch off.

As a general rule, the radar will give less trouble if left running, or, where possible, switched to Stand By, in preference to being frequently stopped and started. If the radar is switched off for a long period, it is recommended that it is set to Stand By periodically. This avoids condensation which could cause damage to the EHT circuits in both the display and scanner units.

Plate 12 Typical console-mounted display

Health precautions and servicing

These are taken together as they are interrelated

Health
The following, which should be found in the *Owner's Manual*, is addressed to him rather than a service engineer, for whom there are further precautions.

(a) High voltage. At several points in the equipment there are high voltages, enough to kill anyone coming into direct contact with them. Even after switching off, some capacitors may take several minutes to discharge. Always set the mains switch fuse to OFF and withdraw the fuses before removing the outer cover of a unit.

(b) Radiation
(i) Users of cardiac pacemakers should be aware that radar and other RF radiations can upset some pacemakers. If you suspect a malfunction, leave the vicinity of the radar system immediately.
(ii) Avoid the possibly harmful effects, especially to the eyes, of radar transmissions by keeping outside the hazard zone of the aerial, or *open* waveguide from which power is being radiated. The hazard zone is calculable from a complicated formula, but is *less* than the radius of a revolving aerial. In practice, therefore, the only danger from a revolving aerial is to be hit by it. Closer approach to a stationary but transmitting aerial is only likely to be made when servicing the radar and the only recommendation to an owner is: don't. However, if it is necessary to work on the scanner unit, always set control to Stand By or OFF.

Which leads to the whole question of . . .

Servicing
Unless he is an electronic engineer – and even then a little knowledge is a dangerous thing – an owner is best advised to leave all servicing to a recognised dealer or service agent. The only exceptions are changing fuses and crystals. Some, particularly mechanical, jobs are not difficult; for instance, adjustment of the centre spot. But this needs the radar to be on and it is the associated hazards such as high voltages in various other parts around that can be dangerous. In short, a radar is not just a souped-up television set, and servicing should be left to professionals.

'R' WA 211° x 1.6 miles
057°
T co. /speed 268° x 8.0 kts

'S' WA speed
Target stationary

'T' WA 284° x 3.4 miles
057°
T co./speed 341° x 17 kts

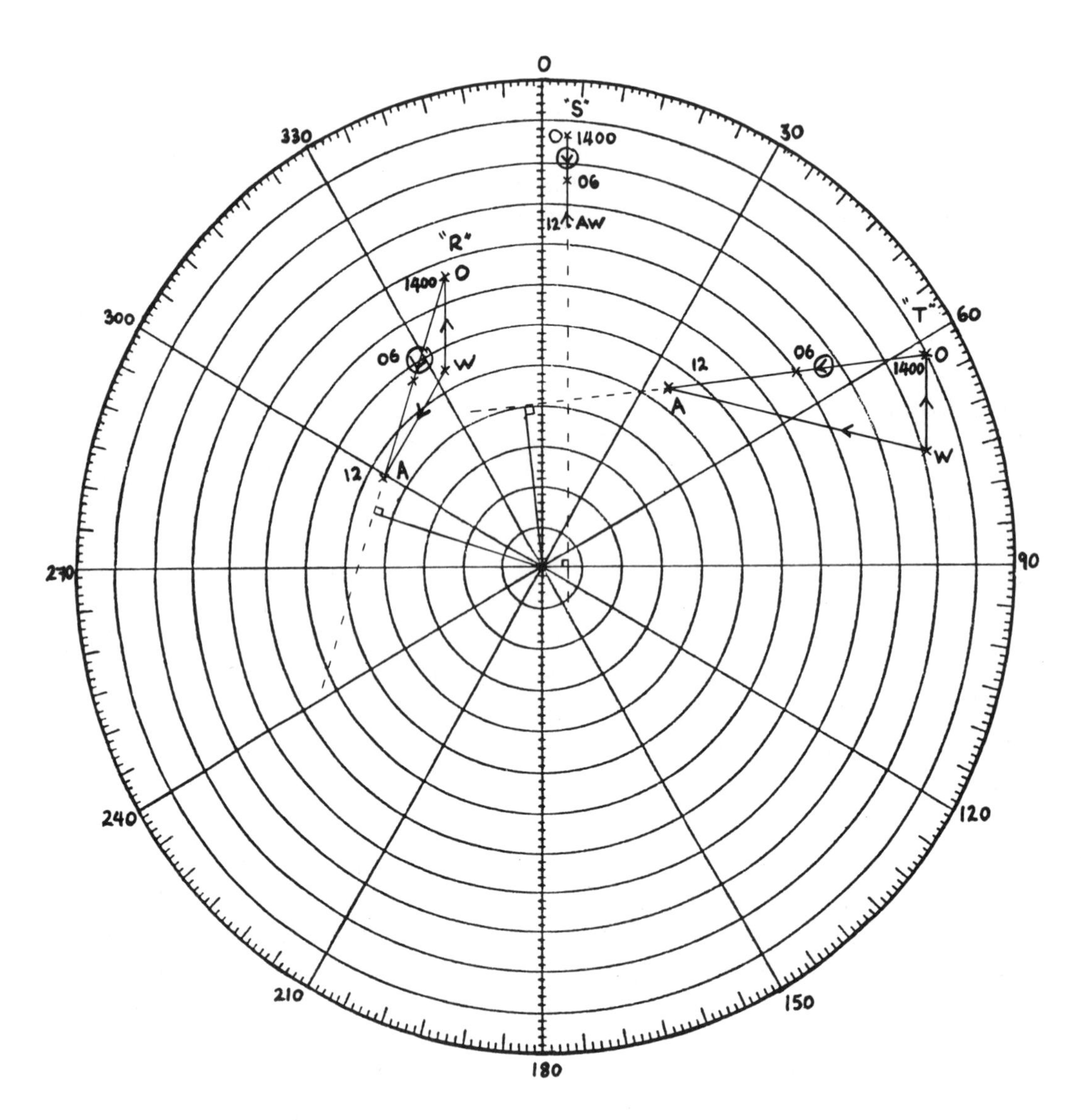

Fig. 10 Spider's web plotting sheet.

Radar Plotting

It was stated earlier, when discussing collision avoidance, that in conditions of bad visibility it was irresponsible to alter course without having first ascertained the other ship's course and speed. Unless you have a £20 000 computerised radar, the only sure way of doing this is to plot. Plotting is not a black art but a straightforward process that can be easily learnt by practising in good visibility, proficiency (and added peace of mind) rewarding the effort.

Plotting has three main purposes:

(a) To ascertain how near a target will pass, at what time, and if risk of collision exists.
(b) To ascertain the true course and speed of a target.
(c) To ascertain the predicted tracks of a target (for indicating when it alters course/speed, etc.).

Team of Two

Whenever possible, two people should be employed on plotting, one at the radar, the other on the plot. When the boat is yawing, closely spaced echoes are easily mixed up. The radar observer should identify each target separately, track it and convey the bearing and distance to the plotter as required.

Spider's Web

The plotter has a choice of basic tools. Either you can use the compass rose of a spare chart or you can buy (or draw) a 'Spider's Web' plotting sheet (Fig. 10). The advantage of the latter is that only a pencil and ruler are required, leading to faster results. If you draw your Spider's Web with a radius of 12 in., you can use 12 range rings. If the 6 mile range is chosen, then each ring would equal half a mile.

When an echo is observed on the radar, the skipper (who may of course be one of the plotting team, though if possible should be free to keep a visual look-out) will want to know a number of things about it: initially whether the bearing is opening, steady or closing and if the range is increasing, steady or decreasing. Plotting will provide this information, and more.

Standard drill

Experience has shown that a recognised drill greatly enhances efficiency. The advantages to the small boat man are: the drill is well tested, other trained people will fall in with it immediately, and nothing will be overlooked.

It is now necessary to explain some definitions.

(a) Bearings to port of the ship's head are described as Red, those to starboard as Green (i.e. Red 8, Green 135).
(b) T = true (meaning actual – not relative – course, speed or bearing).
(c) A bearing is said to be steady if it remains constant; closing if the angle on either bow is decreasing; opening if the angle is increasing.
(d) Closest point of approach (CPA) is the length of a line from own ship that cuts the projected motion of the other ship at right angles.

The object is to make a standard report to the skipper in two parts:

Part 1
(a) Last bearing: drawing forward or aft, passing clear, or steady (i.e. crossing ahead, passing astern, otherwise passing clear; or on a steady bearing – meaning risk of collision if range is decreasing).
(b) Last range: decreasing, increasing or steady.
(c) CPA.
(d) Time of CPA.

If the target is going to pass within, say, 2 miles and especially if avoiding action will have to be taken, Part 2 must be completed:

Part 2
(a) True course of target.
(b) True speed of target.

The following facts should always be borne in mind:

- All target information is gleaned on its past motion; target course and/or speed can change at any time.
- When own boat is *not* making way through the water, target's course is shown *relative to own ship's head but target speed is true.*
- When target vessel is stopped, no heading can be detected (there being no past motion).

Plate 13 Big motor-sailer with 48-mile RM 914 radar at about 24 ft. Target of 80 ft would be recorded at 17 miles

Plotting exercise 1 (Ascertaining other ships' movements)

Now let us start plotting. Assume we are making way heading 057° at 6 kts. The radar is on the 6 mile scale and the following radar observations (see Fig. 10(1)) of three targets R, S and T are recorded (all bearings having been corrected for course error as on page 16):

	Target R			*Target S*			*Target T*		
Time	Bearing	Bng. Change	Dist.	Bearing	Bng. Change	Dist.	Bearing	Bng. Change	Dist.
1400	341	–	3.8	003	–	5.4	062	–	5.5
1406	325	16	2.8	003½	½	4.8	054	8	4.0
1412	299	26	2.3	004	½	4.2	036	18	2.7

Note that time intervals are equal to facilitate evaluation of the various changes, and in multiples of 60 to facilitate speed calculations.

Now, in each case a 'velocity triangle' has to be constructed. A time interval of at least 12 minutes is usually adopted, as here. The three corners of the triangle(s) are always listed O, A and W:

O is the origin – the first plotted position of the target.
A is its position – (in this exercise) after 12 minutes. OA is therefore the movement of the target as seen on the radar, after 12 minutes. OA extended thus . . . is the target's predicted motion.
W is zero speed point. WO is 'way of own', WA is 'way of another' (i.e. the other vessel's course *relative to our heading*, and her true speed).

CPA (Closest Point of Approach) is found by drawing a perpendicular from the centre spot (own ship) to OA extended.

From our plot we can report to the skipper that we have three targets on the 6 mile scale, as follows:

Target R bearing Red 61, opening fast. Range 2.3 miles, decreasing. CPA bearing Red 72, range 2.3 miles, in 2 minutes, at 1414. True course and speed 268° × 8 kts. (Found thus: WA = way of another = 211° Relative. (Own course) 057° + 211° = 268° true. Distance in 12 minutes is 1.6 miles = 8 kts.)

Target S bearing Green 4, just opening. Range 4.2 miles, decreasing. CPA bearing Green 90, range *3 cables*, in 42 minutes, at 1454. Target is stationary – could be a buoy, ship stopped, etc.

Target T bearing Green 36, closing rapidly. Range 2.7 miles, decreasing. CPA bearing Red 7, range 2 miles, in 7 minutes, at 1419. True course and speed 341° × 17 kts.

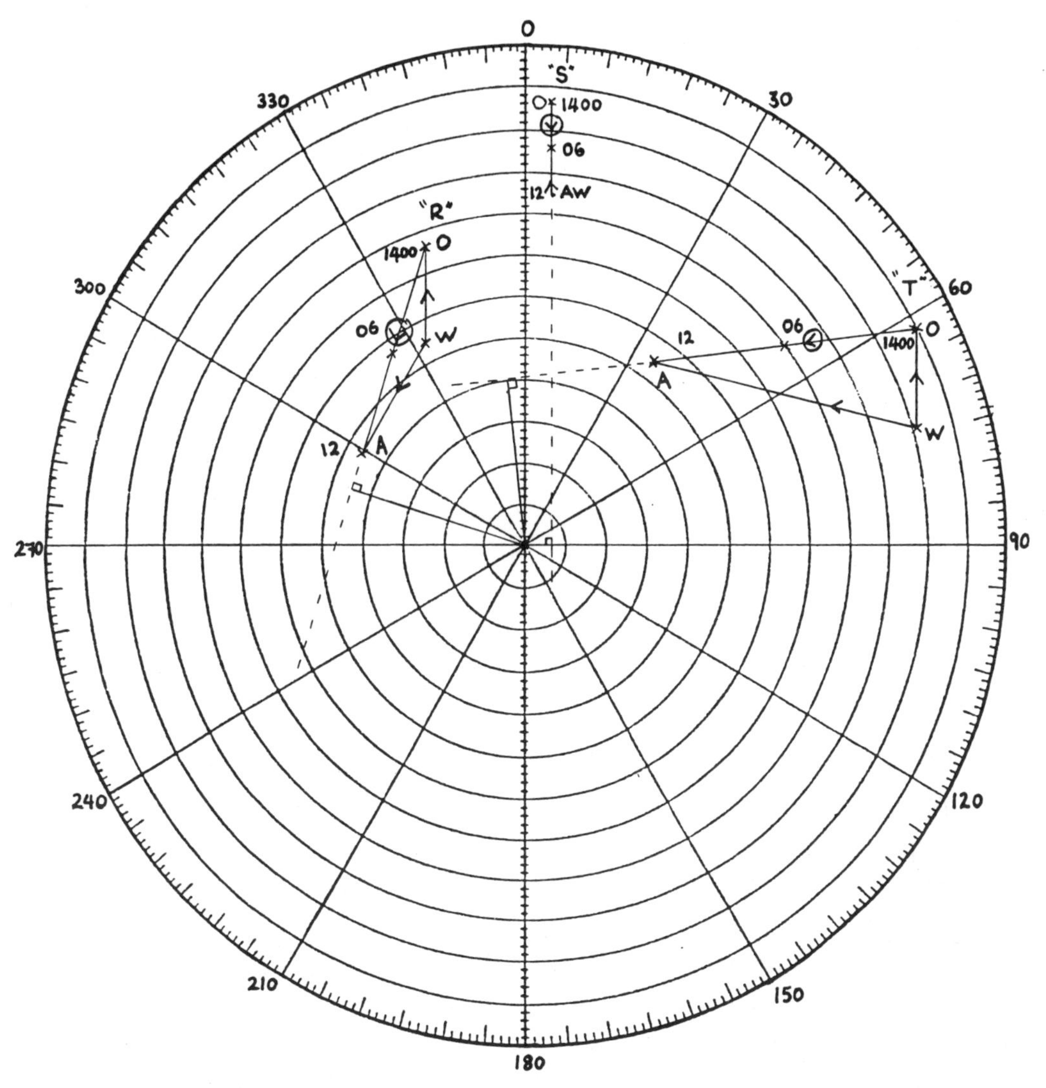

Fig. 10(1) Spider's web – repeat of Fig. 10.

Plotting exercise 2 (Ascertaining other ship's alterations)

In Exercise 1, all three targets were plotted by recording the apparent movement they had made on the PPI. Remember that a vessel can alter course or speed at any time. It is therefore very important that we continue to plot each target to ascertain at once if it is not following its predicted motion.

If we maintain our course and speed, all targets will proceed along OA extended, at the same rate already recorded for OA. If a target does not, however, a new plot will have to be made to ascertain her change of course and/or speed.

If we were to stop at 1412, with our boat still heading 057°T, all targets would proceed along WA extended, at the rate already recorded for WA. This would mean that both targets R and T would increase their CPAs, and target S would remain stationary in the position recorded for 1412, taken that none of the targets alter their course or speed. In the case of target S, it is understood that it remains stopped.

'R' WA 211° x 1.6 miles
057°

T co. /speed 268° x 8.0 kts

'S' WA speed

Target stationary

'T' WA 284° x 3.4 miles
057°

T co./speed 341° x 17 kts

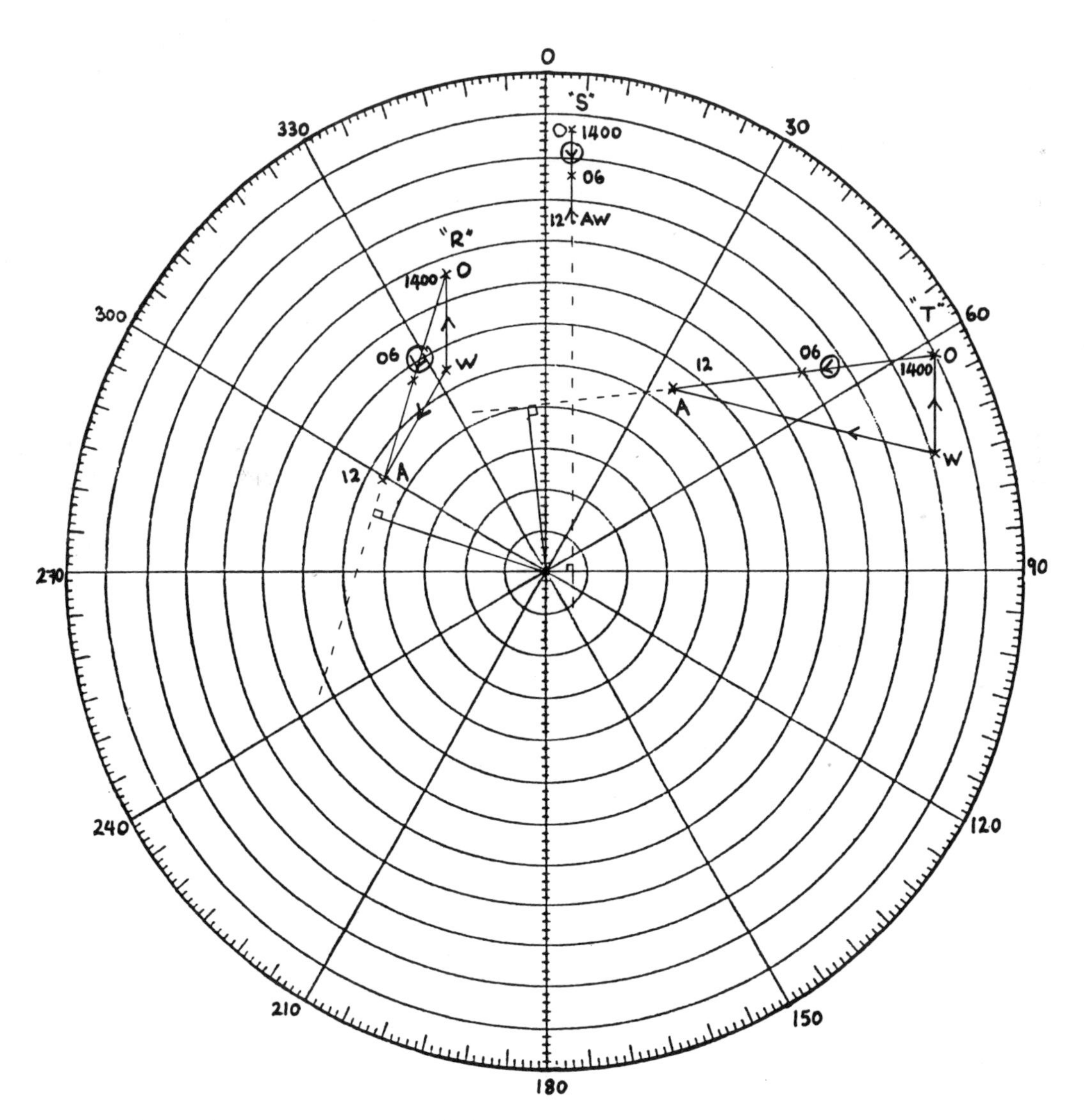

Fig. 10(2) Spider's web – repeat of Fig. 10.

Special interpretations

The foregoing represents the basic groundwork of radar plotting. Further study is recommended, to take in, for instance, rate of change of bearing and plots when own ship and target ship alter course (see Figs 11, 12 and 13). Familiarity with these will provide useful information quickly. Take, for instance: *rate of change of bearing* (i.e. change in a given time). For a passing target, not only does the bearing open, but the *rate* of change of bearing increases, prior to arrival at CPA. This can be seen from the observations of the target that is tracking from A to B (Fig. 11). The bearing is opening and the rate of change of bearing increasing; the target must pass clear. Remember to use equal time intervals for 'rate of change'.

Now study the target that is tracking A to C. Although all the bearings continue to open, the rate of change of bearing remains steady and then decreases (after the target alters course to starboard). These observations, coupled with a decrease in range, should warn the radar plotter that something serious is up and there is a risk of collision.

Figure 12 shows the effect of own ship stopping at 1418 and resuming course and speed at 1454. It is understood that the target ship maintains her course and speed throughout.

Figure 13 shows the effect of own ship altering course 90° to starboard, maintaining speed, at 1418. (WO_1 = new way of own ship. O_1A = new target motion. 18 . . . 30 = new predicted track of target.)

OWA triangles are shown in Fig. 14 (page 44).

With time and practice, any radar plotter can recognise:

(a) Stationary target.
(b) No change in course or speed of target.
(c) Alteration of course by target to starboard.
(d) Alteration of course to starboard and increase in speed by target.
(e) Reduction of speed by target (no alteration of course).

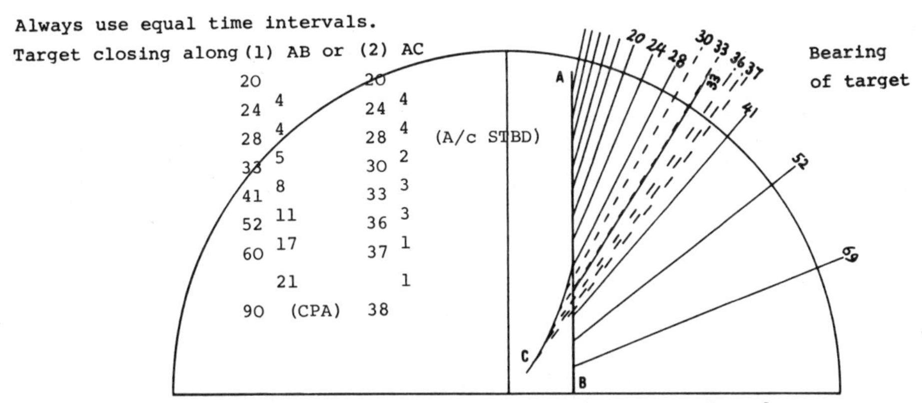

Fig. 11 Information learnt from rate of change of bearing.

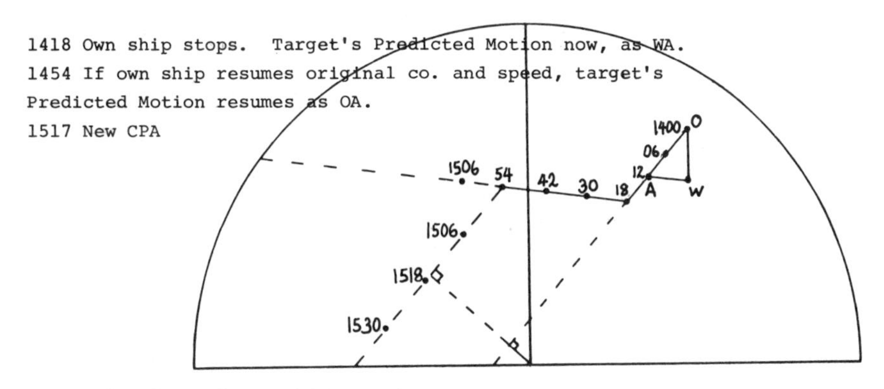

Fig. 12 Effect of own ship stopping.

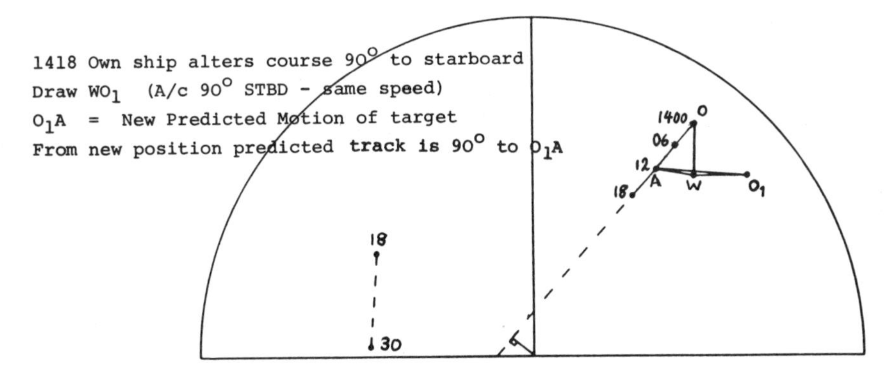

Fig. 13 Effect of own ship altering course.

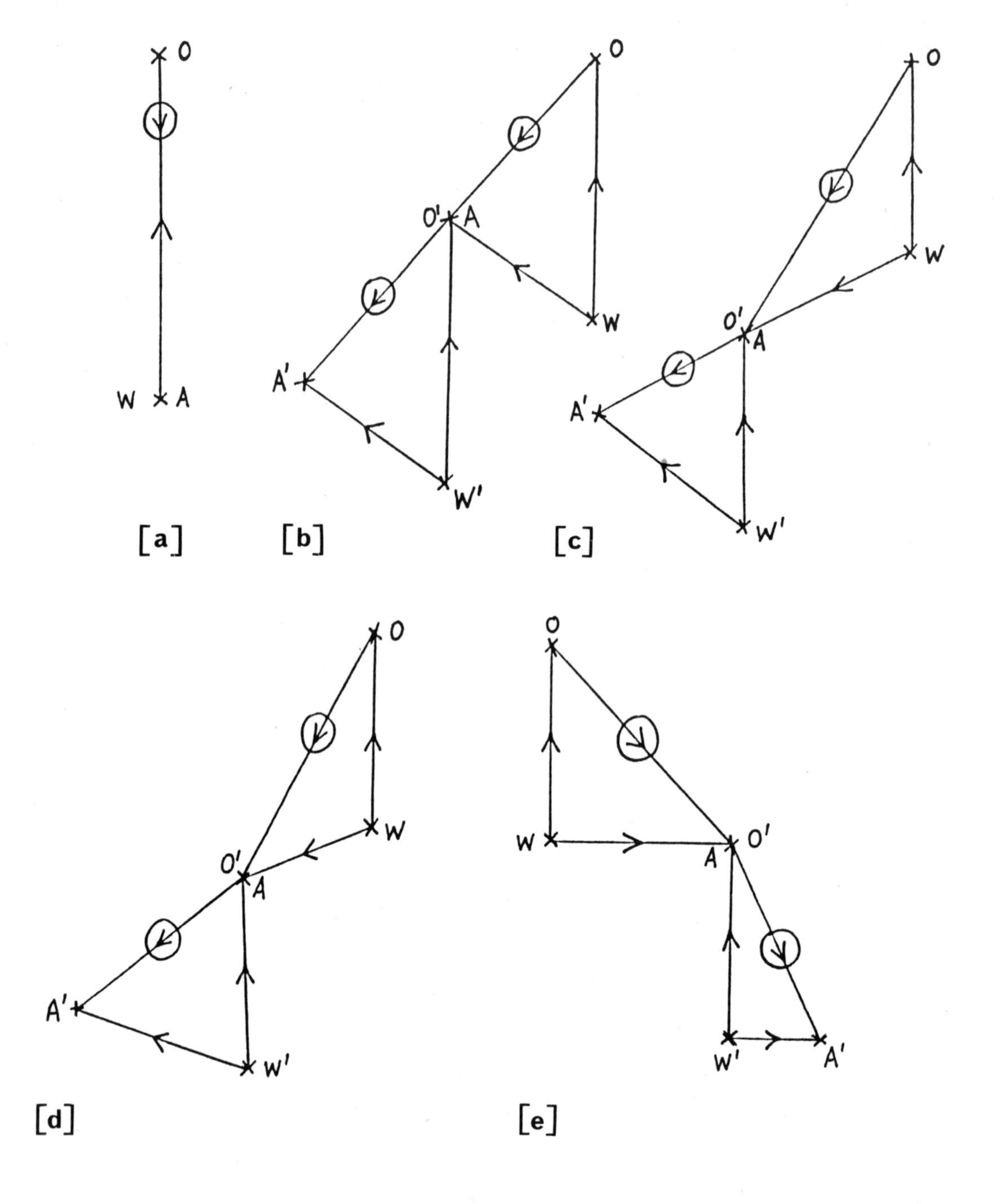

Fig. 14 OWA triangles.

Tips on radar drill to remember

(a) The skipper only knows what you tell him.
(b) He must be told the true course and speed of any target (i) whose bearing is not changing appreciably and whose range is decreasing, and (ii) whose CPA will be 2 miles or less (or whatever distance has been mutually agreed).
(c) Continuity of radar observation by the same observer is essential.
(d) If a feeling of uncertainty or impending risk arises, it is often best to stop the boat and sort out the echoes.
(e) Practise plotting in clear weather, when mistakes may not matter. Only in this way will real proficiency be attained.
(f) Ordinary seamanlike caution dictates that you should not spend too long at a time plotting. Keep the Mark I eyeball in constant use, and ears too, in fog.

Plate 14 Remember the Mark I eyeball!

Blind Pilotage

Blind pilotage is the navigation of a vessel in restricted waters in little or no visibility. In making the decision whether or not to undertake blind pilotage at all, consideration must be given, among other factors, to the state of your radar, the competence of the radar observer, and the safety of the boat and crew in relation to the degree of necessity to press on. It should be remembered that (a) radars have been known to break down, and (b) many a competent professional mariner has come to grief in reduced visibility.
(See section on Rule of the Road on page 14)

Radar errors

Remember that the range facility is accurate, but bearing not so (1° of error at 1 mile is 33 yards and at 6 miles is 200 yards). Some other points to check are that the centre spot is in fact in the centre, that the heading marker is lined up with the aerial, that range rings are equidistant and that they correspond with the variable range marker, if fitted. Your handbook will explain how to adjust the heading marker, but the others are a matter for the service engineer.

Plotting errors

These can be of bearing, range, course and/or speed of own ship, and time of plotting intervals.

Remember that (a) errors in range, bearing and own speed = large error in course and speed of target, if her speed is slow compared to yours; (b) the slower the speed of the target vessel, the more unreliable the plot; and (c) doubling the plotting interval will halve the chance of error in CPA.

Extra observer

If a fix is going to be obtained via the radar every 6 minutes, the plotter is going to be fairly busy. Since the radar is the only means of detecting moving objects around you, it may well be prudent to station another observer, if available, particularly to advise of any moving targets. With precise training and timing, contacts can be plotted every 3 minutes, without interfering with the 'position' plotter.

Timing is of paramount importance for accurate results. For example, if you are taking bearings every 3 minutes and are 25 seconds slow, this represents an error of 13.8%.

Whenever in blind pilotage conditions, a prudent observer will continuously:

(a) Take every opportunity to cross-check radar with visual bearings.
(b) Ensure radar is following ship's head correctly.
(c) Ensure that all times are taken from one clock close to the radar.

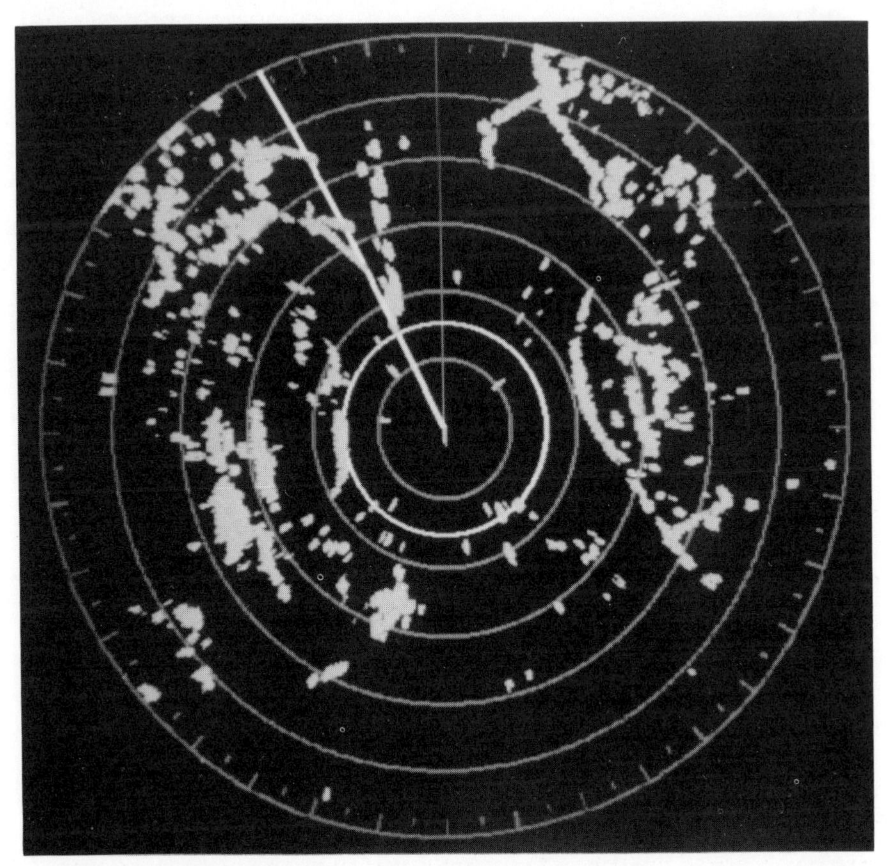

Plate 15 Check that the centre spot is in centre, heading marker is lined up, range rings are equidistant . . .

Blind pilotage exercise

The basis of blind pilotage is to track parallel to an unmistakable radar reference line, at a given distance from it, the pre-determined reference line being easily established on the PPI.

For exercise, let us pre-plan a blind pilotage from W. Bramble buoy to Hamble Point buoy, in the Port of Southampton. Our draught is not more than 6 ft. The exercise can be followed on Admiralty Chart 1905 (see Fig. 15). A track outside of the Main Channel is desired.

(a) Draw reference line AB, joining E'ly points of land at Stansore Point and Calshot Spit (036/216).
(b) Draw track parallel to AB, 1.03 miles East of W. Bramble, to clip Calshot buoy to the East, and extend to 2 metre contour line.
(c) Draw reference line CD, joining N'ly point of Calshot Spit, and S'ly end of Fawley jetty (144/324).
(d) Draw track parallel to CD, from two cables North of Hamble buoy, back to cut original track. Where they cut is your alter course position, which is 1.18 *miles* from Meon coastline. (1 cable = 200 yards.)

You now have two tracks, with parallel radar reference lines. On your first track, your reference line distance is 1.03 miles to port. You alter course when the land dead ahead is 1.18 miles. On your second track, your extended reference line distance is 4.7 cables to port.

That is your basic blind pilotage, with a principal radar range of 3 miles. Obviously, leeway, tidal effect, etc. have to be included, as in normal navigation, to ensure you keep on track.

To perfect the pre-plan, aids and reference positions are clearly shown on the chart. Try for as near 90° cuts to track as possible.

Exercise extreme caution when crossing the North Channel (2 cables). If possible keep listening watch on VHF 12 (working channel of Port Radio) and set echo sounder alarm to 7 ft. It will not read less if track is followed.

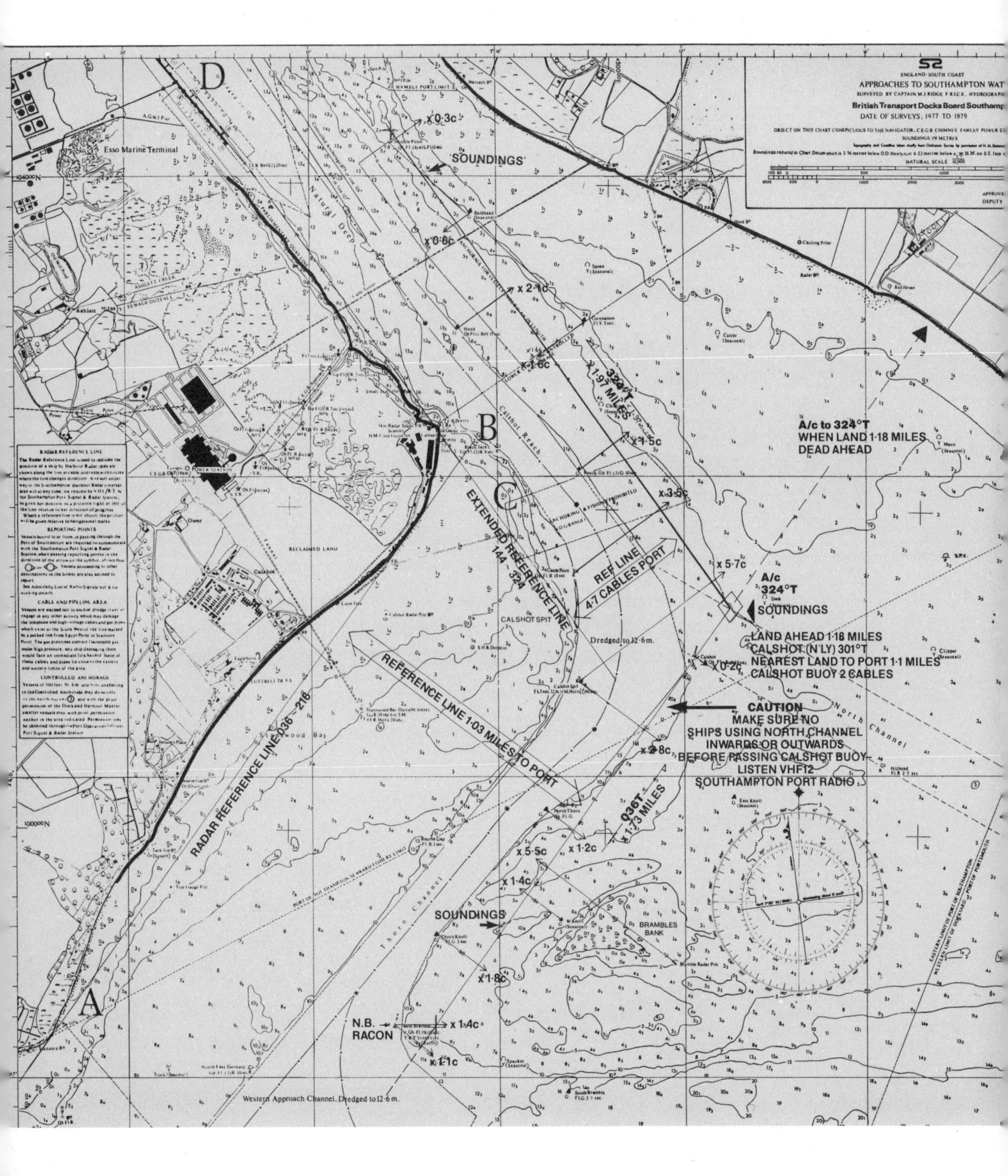

Fig. 15 Pre-planned blind pilotage from W. Bramble buoy to Hamble point buoy.

Tests

The persevering reader may like to take a couple of tests. The first is an aptitude test which should be correctly answered in 2 minutes, 20 seconds. The second is based on the information on plotting provided. This should be completed in 6 minutes and 40 seconds.

Test 1 Aptitude Test (Time 2 minutes, 20 seconds.)

(1) Your speed is 12 kts. How far do you go in 6 minutes?
(2) Your radar plot indicates that a target has travelled a distance of 2.3 miles in 10 minutes. What speed is she making good?
(3) What is 0.3 × 0.2?
(4) You are steering 120° true. A target vessel is plotted as steering a relative course of 350°. What is the true course of the target vessel?
(5) Whilst steering a course of 130° true, a target bears RED 20 on the radar. You alter course 90° to starboard. What is the bearing of the target on the radar now?

Test 2 Questions on Radar Plotting. (Time 6 minutes, 40 seconds.)

(1) You are trying to make good a course of 093° true, but you are yawing heavily. The radar observer calls out the following five relative bearings. As he does so, you record the ship's head at the time. What are the correct relative bearings you would use when plotting:

Bearing	Ship's head
013	096
093	099
351	095
342	086
359	090

(2) As the radar plotter, explain the correct format, in order, of how to report a target in fog to the skipper.
(3) As a radar plotter, you know the only way to ascertain what a target is actually doing is by constructing a velocity triangle.
    (a) What are the letters used at the points of the triangle? Write down the three sides of the triangle, and explain what each side represents. (Answers (b), (c), and (d).)

(e) Which point of the triangle is especially important, and what do we call this point?

(4) What are the four main plotting errors?

(5) What action can you take that will halve the chance of errors in plotting the CPA of a target?

The answers are given below.

ANSWERS TO TEST 2

(1) Correct relative bearings are: 016, 099, 353, 335, 356.

(2) Last bearing: Opening, closing, or steady.
Last range: Increasing, decreasing, or steady.
CPA: Relative bearing, and range.
Time of CPA.
True course of target.
True speed of target.

(3) (a) OAW. (b) WO = Way of own ship. (c) WA = Way of another ship. (d) OA = Target motion. (e) W = Zero speed point.

(4) (i) Errors in bearings. (ii) Errors in range. (iii) Wrong estimation of own ship's course and/or distance during plotting interval. (iv) Errors in time of plotting interval.

(5) Doubling the plotting interval.

ANSWERS TO TEST 1

(1) 1.2 miles $(6/60 \times 12 = 1/10 \times 12 = 1.2)$

(2) 13.8 kts $(60/10 \times 2.3 = 6 \times 2.3 = 13.8)$

(3) 0.06 $(2 \times 0.3 = 0.6)$

(4) 110°T $(350 + 120 = 470 - 360 = 110)$

(5) RED 110. $(20 + 90 = 110)$

Plate 16 Practise plotting in clear weather

Appendix I
How Radar Works

Basic principle

The basic principle of radar is simplicity itself: if radio energy at high frequency (usually between 1 GHz and 40 GHz) is emitted with high power, it will be reflected off hard objects in much the same way as audible sound. Since the speed of both the emitted and reflected energy is known, it is only necessary to construct an electronic timing device for the range of the object returning an echo to be measurable. Its bearing is even more easily determined of which more later.

In practice a radar system consists of a transmitter to generate pulses of radio-frequency energy; a rotating aerial for sending these out into space in the form of a narrow beam; a receiver for accepting the echoes returned (via the aerial) from any suitable target; and a display for presenting them visually so that range and bearing are evident. There must also be a power supply.

Transmitter

Ranging by radar is very similar to making use of echoes of the human voice. If one shouts in order to time an echo, the best results are obtained if 'transmission' consists of short, sharp sounds. The shortness ensures that the transmission will have ceased before the echoes return and the sudden rise to full power will make the instant of return immediately perceptible. Similarly, the radar transmission which must operate for ranges varying from 25 yards to, say, 48 miles, has to produce a short pulse of oscillation rising rapidly to full amplitude, which it maintains until cut off. The duration of the pulse is called the pulse length, and its frequency, the pulse repetition frequency (prf).

If, for instance, the target is only 50 yards from the transmitter, the pulse will have to be cut off before the beginning of the wave has had time to travel to the target and back; a total distance of 100 yards. A radio wave travels 328 yards in one microsecond, so the pulse, in this case, must not be longer than 0.3 microseconds. If it is also desired to receive echoes from targets up to, say, a 30-mile range, the interval between pulses must be long enough to enable the wave to travel twice this distance, i.e. 370.4 microseconds. This gives a maximum prf of 2 700 pulses per second. (The pulse has to be further shortened in order to separate echoes of the same bearing and of very small difference in range. With too long a pulse, the echoes would be merged.) In practice a choice of two or more prfs are provided for use at different ranges.

Operation

It will be noticed that in the case mentioned the transmitter is required to oscillate for 0.3 microseconds and then to rest for 370.1 microseconds. Because of this, a small valve (the magnetron) may be used to generate very high power since it has relatively long intervals in which to cool. In practice the prf is usually between 500 and 4 000 pulses per second. In these circumstances a magnetron no bigger than a 250-watt lamp can give a peak power of 60 kW, though 5 kW is typical for yacht radars.

The 'firing' of the magnetron (an apt description) is the last action from a chain of four basic circuits in the transmitter. Traditionally (in the latest solid-state equipment there is some variation) these, and their functions, are:

(1) The sync. trigger, which synchronises the display with the transmitter.
(2) The sub-modulator, which is a timing device determining pulse width.
(3) The modulator, which stores energy and then releases it to the magnetron on command from the sub-modulator.
(4) The magnetron itself, which converts electrical energy into electro-magnetic pulses.

These circuits are so designed that, after the sudden initial rise, the discharge is as near as possible at a constant rate and is completed in the time of the desired pulse length; that is to say, it takes the form of a short, square pulse. The magnetron bursts into oscillation when the energy is released and ceases to oscillate when the supply is cut off. Though it has only two electrodes, the action of the magnetron is complex. Suffice it to say that it is known as a cavity resonator and operates between the poles of a very strong magnet.

As we have seen, the key to range measurement by radar is the time interval between transmitted pulse and returning echo. To measure this, the sync. trigger sets off the timing arrangements in the display at the same time as the magnetron is fired and the radio frequency (rf) pulse begins. We will return later to the display.

The rf pulse so produced passes from the transmitter to the aerial via a wave guide, which is simply a tube (usually of copper) of rectangular section, engineered to high limits. In yacht radars it is only a few inches in length.

The aerial

There have been several types of aerial over the years but the most usual is the slotted waveguide design. This basically consists of a length of waveguide mounted horizontally on a turntable so that it can be revolved. The front face has a series of slots cut in it, again engineered to very high limits in both pitch, angle and depth. The aerial has an electric motor which drives it at about 30 rpm, the whole assembly being known as the scanner. The beam radiated is narrow in the horizontal plane – 4° or less – an important factor in bearing discrimination (the ability to separate adjacent echoes). The vertical beamwidth is about 27°, to allow for the roll of the ship and for very close targets.

Plate 17 A variety of scanner (aerial) installations

The advantages of the slotted waveguide design (two other designs being the tilted parabolic cylinder and the cheese) are in the realms of weight, wind resistance and sidelobe performance. Construction is, however, expensive and the other types are sometimes found in the cheaper small boat equipment. Double aerials exist where one is used for transmission and the other for reception. Usually, however, the reflected energy is made to return through the same aerial and is passed down the waveguide to the receiver.

The receiver

The transmitted pulse is extremely powerful, as we have seen, and the returning pulse naturally very small (much of it having been scattered in unsuitable directions); so the sensitive receiver must be protected from the transmitted pulse. The usual protective method is to use a gas-filled transmit/receive cell, or solid state limiter, which blocks the transmitter pulse from the receiver input.

The receiver's function is to amplify the minute returning echo pulses while retaining their distinctive shape so that they will be capable of giving optimum resolution on the display. Since it is expensive to amplify the returning pulses at their own radio frequency (e.g. 9 MHz), the first duty of the receiver is to change the frequency to a more suitable one. This change is effected in the mixer, the main components of which are the local oscillator and the mixer diodes. The principle is the same as in the superheterodyne radio receiver, i.e. mixing of the received signal with one of different frequency produces another of intermediate (difference) frequency; this, after suitable amplification, is detected to produce video pulses suitable for acceptance by the video amplifier of the display. A major problem is in keeping down the level of 'noise' generated by the receiver mixer and input circuits. Noise appears as a speckled background on the radar display and must be kept to a minimum if weak, legitimate echoes are to be seen.

The display

The display is the name given here to the entire unit, including PPI, electronics and controls. The PPI is the end of a cathode ray tube (CRT) which (with relative motion) presents the same picture of the area around the transmitting aerial as would be seen by a helicopter flying so as to remain always immediately over own ship. The object of the display designer is thus to provide echoes on the PPI which are in the correct positions relative to the centre spot, which represents own ship.

Taking range first, consider the situation with the scanner stopped. It is sending out pulses and receiving echoes from another ship. Now the mangetron's trigger also starts a spot on the PPI moving out from own ship to the circumference at a predetermined speed. (This spot is actually moving at such a rate that it appears as a solid line, known as the timebase or 'trace', in practice dimmed to a barely visible level.) When the receiver picks up the returning echo, it causes the spot to increase in brightness. Because the spot is travelling at a predetermined speed, its distance from the centre when brightened can be measured in time. This time is the same as that taken for the radar pulse to go out and its echo to be received. We know the speed of the pulse and so we can assess the distance away of the ship returning the echoes.

In fact, to paint the echo on the PPI, the momentary increase in spot brightness calls for variation of one of the voltages applied to the CRT. For the echo pulse to provide this variation, it must (after being processed in the receiver as described) receive a final amplification by the video amplifier. The video amplifier is also responsible for the painting of measuring marks on the PPI.

Still considering range only, a choice of scales is provided so that at the turn of a control, the radius of the PPI can be altered to represent, say, 24, 12, 6, 3, 1½ or ¼ mile. (As implied earlier, it is often necessary to vary pulse repetition frequency with range and this is effected automatically by the same control.) If the

Plate 18 Display of the RD 130 radar (this is monochrome DSC)

display is on the 6 mile scale and a ship 3 miles off, the echo will of course appear exactly half-way to the circumference.

To facilitate judgement, range rings are provided at set intervals. If an echo exactly cuts a ring, its range is clear, but if not, interpolation is necessary by eye or with a ruler (except where a variable range marker is fitted; this provides the answer in digits on completion of a simple measuring action).

Bearing

What of bearing? The receiver will only register an echo when the aerial is pointing at a target. Therefore, if we rotate the trace in the display exactly in synchronisation with the aerial, the spot going out along the trace will be momentarily brightened at the exact moment the aerial is pointed at a target; and therefore the echo will show up on exactly the right bearing.

In practice the face of the CRT is coated with phosphor which ensures that the echo continues to glow after the trace has passed, and it can be arranged that all targets within range glow all the time, being renewed in brightness at each revolution of the trace. A bearing cursor is provided (a sheet of glass or Perspex over the face of the PPI) which is revolved until a line thereon – the bearing marker – cuts the echo in question. The bearing can then be read off the scale provided.

The bearing will of course be purely relative to the ship's head. (As with the variable range marker, in some cases the bearing may appear digitised in a 'window' as an electronic bearing marker is aligned.)

Actual units

In practice, the transmitter and receiver are combined in one unit called the transceiver. This is located immediately under the scanner (improving performance by obviating the waveguide which in large radars joins the two). The 'top unit' so formed can include the power supply but this is usually in the display. Thus the five components of a radar system are ingeniously grouped – in small craft radars – to form only two separate units, the scanner (or top unit) and the display, the two being jointed by a co-axial cable. This greatly facilitates installation.

Appendix II
Glossary of Terms

Aerial/Antenna: Part of system from which radio waves are transmitted and their echoes received (see *Scanner*).

Afterglow: The slowly dying glow of echoes after passage of the trace.

Anti-clutter control: A control of the receiver circuit to reduce intensity of echoes created by wave action, or sea-return.

Attenuation: Weakening of radio wave power due to absorption or scattering.

Azimuth: Bearing in degrees measured in a clockwise direction.

Beamwidth: Angular width, horizontal and vertical, of the radiated pulses.

Bearing marker: Line bisecting the bearing cursor, for alignment on an echo to be measured. Can be electronic bearing line (EBL).

Blind arc: Sector on a PPI in which echoes will not be seen due to intervening structures, such as mast.

Cathode ray tube (CRT): The display tube on which echoes are painted, with own ship at centre (relative motion presentation).

Cavity magnetron: Valve in the transmitter which produces radio pulses at high power for emission by the scanner.

Clutter (rain): Echoes created by rain or snow which tend to obscure weak 'legitimate' echoes. See differentiation.

Clutter (sea): Echoes from waves, out to about two miles. Most evident to windward. See Anti-clutter control.

Co-axial cable: Flexible cable to conduct radio frequency energy between the transceiver and aerial.

Compass bearing: Bearing shown by the compass. Deviation must be applied to obtain magnetic bearing (plus variation for true bearing).

Crystal: A synthetic element to control the frequency of the radio pulses.

Cursor: A manually-rotated transparent disc over the display to determine the relative bearing of an echo (see *Bearing marker*).

Definition: A measure of the presentation of echoes on the screen; sharpness or detail.

Differentiation (FTC): A circuit in the receiver to reduce clutter from rain, snow, etc; improves definition at short ranges.

Discrimination: Ability to separate the echoes from two objects that are close together in range or bearing.

Display: Loose term for PPI, but also applied to entire unit (CRT, controls, circuits and usually power supply).

Duct: Stratum created by atmospheric conditions that tends to confine radio waves within it; ducting extends range beyond the normal.

Echo (Paint): Spot shown on the CRT created by pulses reflected from an object.

Electron gun: That part of the CRT which emits a stream of electrons to the fluorescent-coated inner face of the tube. Intensity is increased to paint a spot when an echo is received.

Fluorescence: Light emitted from the CRT as the result of a high charge of electrons striking the tube face (see *Afterglow* and *Persistence*).

Focusing: Concentrating the electron stream to produce a sharply defined spot on screen.

Gain: The ratio between input and output power; analogous to the volume control of a radio.

Heading: The direction in which a boat is pointing, the magnetic course being steered.

Heading marker: Electronically-generated line on PPI indicating the heading.

Interference: Intermittent and random paints generally created by transmissions on the same frequency by other ships' radars.

Ionosphere: A layer of heavily ionised molecules in the outer part of the earth's atmosphere, far beyond the stratosphere.

Lobe: The volume of space in which the radiated pulses are concentrated.

Magnetron: See *Cavity magnetron.*

Microsecond: One millionth of one second.

Mixer: The part of the receiver which changes the frequency of the received radio pulses to a frequency that permits amplification of the weakest signals to a level that will show on the CRT.

Multiple echoes: When two vessels are passing on opposite courses at close range and are beam to beam, a second echo may appear beyond the normal one (or a series of echoes at equal intervals).

Persistence: Degree of afterglow.

Plan position indicator (PPI): Radar picture presented on CRT as if observer was in a helicopter hovering over own ship. (Relative motion.)

Power consumption: Amount of electric power required to operate the radar.

Pulse: Regular but intermittent emissions of radio frequency energy; bursts of energy, as opposed to a continuous stream.

Pulse length: Length of time, measured in millionths of one second, for which a pulse is transmitted.

Pulse repetition frequency (prf): The number of times per second that pulses are transmitted (also pulse repetition rate).

Racon: Radar beacon. Transponder ashore or on lightvessel triggered by ship's radar to provide coded signal on PPI indicating range/bearing. (Raymark is a transmitting beacon giving bearing only, like D/F.)

Radar: Contraction of RAdio Detection And Ranging.

Radome: A glass fibre housing used with very small radars to enclose the scanner and other elements of the system.

Range rings: Concentric, electronically-painted rings on PPI for measurement of range from centre spot. Choice of several scales.

Reflector: Highly recommended safety device that can be hoisted on a boat, usually consisting of metal planes, that greatly enhances her normal radar echo.

Relative bearing: Bearing in degrees to port or starboard of the boat's heading.

Relative motion: Presentation where position of own ship, i.e. trace origin, is fixed and all echoes move relative to own ship (see PPI).

Resolution: See *Discrimination.*

Scanner: Aerial plus its rotation mechanism. Also used to denote entire top unit, which may include transceiver.

Scattering: Indiscriminate re-radiation of incident waves.

Scope: Alternative (US) term for 'display', 'screen', 'tube', 'PPI'.

Screen: Fluorescent face of the CRT.

Second-trace echo: Echo which returns to the radar during the period of the trace following that which caused the echo. (Range must be greater than the distance corresponding to the pulse interval.)

Sensitivity: The measure of a radar receiver's capability of detecting weak signals.

Shadow: See *Blind arc.*

Sidelobe: Beams of radio energy radiated on either side of main lobe; relatively low in power, but undesirable.

Signal: Reflected radio pulses or echoes.

Slotted waveguide: Method of construction of aerial in which apertures have been cut for emission of pulses and collection of the returning echoes. Almost standard.

Spoking: Radial, straight line interference caused by a fault in own radar (sometimes faulty contacts on rotating mechanism).

Target: Echo, usually from ship.

Trace: Revolving line across the CRT, synchronised with rotation of the scanner, on which echoes appear as bright spots in relation to their distance from own ship.

Transponder: Unit which, in response to pulses received from a radar, transmits a sequence of pulses which can be recognised by the interrogator (see *Racon*).

True motion: Presentation in which own ship, i.e. the trace origin, moves across the PPI (via inputs from gyro compass and log) with her true course and speed. All echoes of other objects necessarily follow suit.

Variable range marker (VRM): Variable range ring which is expanded or contracted to cut the echo to be measured, its range being displayed digitally.

Waveguide: A hollow metal duct, formed to precise dimensions, through which radio pulses travel to and from the transceiver and aerial. Only a few inches long in most yacht radars.

Appendix III
Radar Range

In standard atmospheric conditions there are three basic 'horizons': geometric, optical and radar. Geometric is the straight line between the sighting point and the horizon; with a point 15 ft high the geometric horizon will be approximately 4 miles distant. The bending action of light waves extends the optical horizon in this case by 6% to 4.25 miles, and the even greater curvature of radio waves extends it by 15% to 4.6 miles. Certain atmospheric conditions will increase the radar horizon even further. (See page 27.)

The radar beam will of course travel over the horizon (the smallest radars' potential being about 14 miles) and the range at which a radar will pick up a target over the horizon is the horizon of the radar plus the horizon of the target. (Range in miles $= 1.22 \times (\sqrt{h_r} + \sqrt{h_t})$, where h_r is the height of the radar aerial and h_t the height of the target, in ft.)

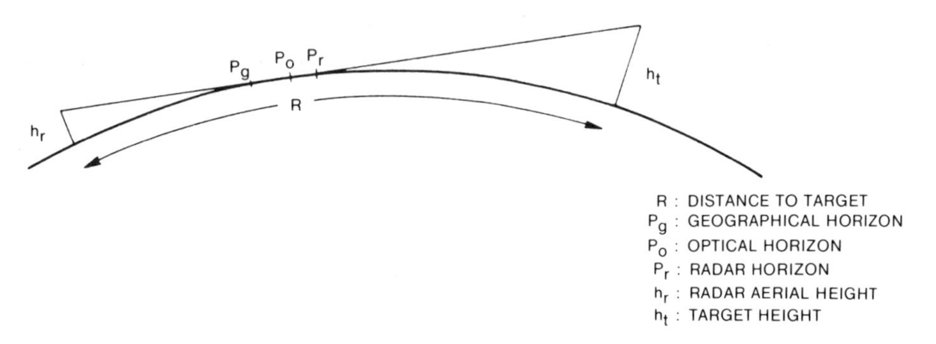

Fig. 2(1) True target distance using radar horizon – repeat of Fig. 2.

The following table gives radar/target horizons:

Height (ft)	*Distance (miles)*	*Height (ft)*	*Distance (miles)*	*Height (ft)*	*Distance (miles)*
5	2½	55	9	110	13
10	4	60	9½	130	14
15	4½	65	10	150	15
20	5½	70	10	170	16
25	6	75	10½	190	17
30	6½	80	11	215	18
35	7	85	11	240	19
40	7½	90	11½	265	20
45	8	95	12	320	22
50	8½	100	12	380	24

It will be seen, for example, that with a scanner height of 15 ft, a 70 ft target will be detected at 4½ + 10 = 14½ miles. It will also be clear that increasing the scanner's height does not increase the radar horizon much: in fact, the height of the target is a more important factor in practice. Gibraltar has been picked up at 86 miles, though this probably brought in other relevant parameters (good reflecting properties, very powerful radar and possibly abnormal atmospheric conditions). See Fig. 2(1).

Appendix IV
Digital Scan Conversion

This is a very short description of the digital scan conversion (DSC) technique as used in the Racal-Decca 370 series Bright Track (BT) small craft colour displays and the larger 970 BT displays.

Conventional radars operate on the Rθ basis, when R is range and θ is the bearing relative to the ship's head. Thus an echo might be at 034°, 6 miles. DSC displays produce an echo at the same spot on the PPI by dividing this up into thousands of minute squares, known as pixels, and then deciding which ones should be illuminated by counting them on the x,y co-ordinates principle: x divides the PPI into vertical lines, counting from left to right, y into horizontal lines from top to bottom, so providing a matrix of cells. There are 484 lines in both directions, enabling a standard TV format to be used, and so permitting the use of recognised and proven TV components. The result is approximately 234, 256 pixels over the whole display, 180 000 on the circular PPI itself. The echo mentioned above might therefore be described as x 150, y 228 in DSC language.

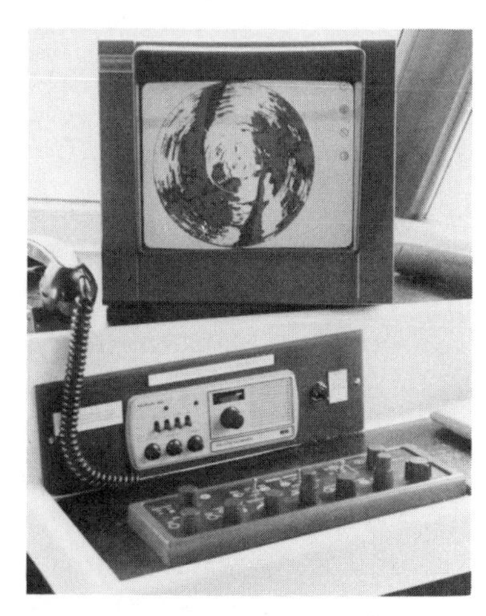

Plate 19 'Bright track' DSC colour display

A pixel either indicates a return or it does not; there are no half measures with only part of a pixel being illuminated. The effect of this is to slightly 'square up' an echo that would normally appear round, but the effect is only noticeable, and then very slightly, where a straight line like the bearing marker is concerned. Close scrutiny shows that this is slightly 'stepped'. Use of a large number of lines would obviate this, but in small displays the miniscule practical advantage would be greatly outweighed by expense, and an optimum is struck. With big displays a larger number of lines is required.

The technique of changing the Rθ-type radar information to *x,y* co-ordinates is known as 'digital scan conversion'. Being an extra step in the radar process and a somewhat complicated one, it naturally adds to the cost of the radar. Let us look at the sequence of events, in an admittedly over-simplified way.

The basic radar information comes down from the scanner in the usual (Rθ) manner. The scanner rotates at 25 rpm, refreshing the picture every 2.4 seconds, but using a 525-line TV-type screen the picture is eventually presented to the observer refreshed 30 times a second. This is what provides both the extra brightness and a steady picture.

The first step in handling the difference in speed is by a technique known as retiming, employed to avoid the necessity of having very high speed circuitry throughout. Incoming video from the receiver is digitised in real time, written into a fast store and then read out at a rate which can be handled by subsequent circuitry. Typically, on a 0.25 mile range, video is input in approximately 3 microseconds and output in 350 microseconds.

Video for two consecutive timebase sweeps is stored in this way and during the readout cycle, only information which correlates on both sweeps is allowed through to the next part of the circuit. This has the effect of rejecting a large percentage of interference (from other radars). It is also beneficial to get rid of spurious signals due to sea and rain clutter before digital scan conversion takes place and, in the case of the 970 BT, the incoming video is next processed by the unique Racal-Decca 'Clearscan' circuitry which continuously and automatically adapts the level of gain to suit the incoming returns.

The next step is to convert the Rθ radar information into *x,y* format which can be displayed as a TV picture. The Rθ to *x,y* scan converter circuit continuously calculates the TV memory addresses as the radar scans in Rθ, and decides where each piece of retimed radar information should be stored in the TV memory matrix. In the 970 BT there are four (three in 370 series) matrix memories, each one being 484 by 484 used pixels, superimposed on each other. These four matrix 'planes' are required to enable not only current radar information but past his-

Plate 20 A typical motor-cruiser, showing the scanner of her RD 370 Bright Track colour radar

tory of targets (track), guard zone information, synthetics such as bearing scale, range rings, etc., and background to be stored and displayed in different colours. The information from the stores is read out 30 times a second and, as we have seen, it is this that provides both the extra brightness and a steady picture. In fact with the Racal-Decca Bright Track it is possible to just see the trace rotating, as it were behind the picture; this reassures the observer that the set is functioning properly.

If this was all, it would be simple to build up the TV picture line by line but in the BT we have to take track history carefully into account. This is done by a 'decrementing' process in which the previous position of the echo is remembered for a certain length of time and then discarded. The remembered part is then displayed, black tracks or green tracks depending on whether it is day or night presentation. The length of time it is displayed is controlled by a programmable memory and optimised for each range scale. The length of tracks thus formed is proportional to the target's relative speed; true speed can be approximated with practice and experience of plotting. Being held in the memory, the BT tracks can be switched on and off. The afterglow smearing, associated with a change of course on the CRT of a conventional display, is obviated, the afterglow in this case being, as we have seen, electronically-generated.

The bearing scale, range rings and heading marker are all stored in the main memory. The VRM and EBL (electronic bearing line) are calculated by a microprocessor. The guardzone uses a correlating technique from scan to scan and from prf to prf. The 'switched mode' power supply operates at high frequency and so is mercifully inaudible.

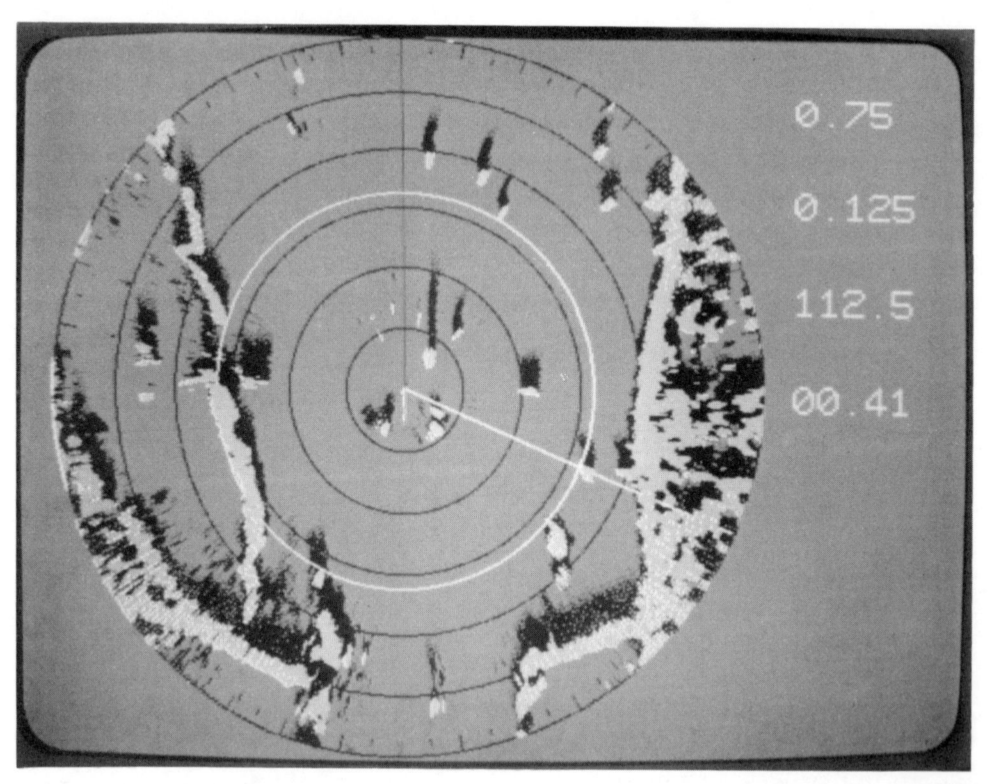

Plate 21 PPI of an RD Bright Track (DSC) colour radar. Note echo tracks proportional to relative speed. Long one on starboard bow denotes a hovercraft. White line and circle are EBL and VRM respectively. Figures at right are: range scale, range rings, bearing, distance

Further advantages, in addition to those listed on page 11, are: should an echo disappear momentarily in clutter its track remains discernible, indicating approximately where it is; instead of the conventional phosphor tube, TV tubes are employed, which are very hard to burn; and the unique way of using colour in the Racal-Decca BT radar is vastly superior to the one in which colour indicates intensity of echo (the latter can be dangerous, as you might have a supertanker opening astern in 'full alert' red, but a small wooden boat right ahead in 'least attention' green).

Index